The Terrestrial Biosphere
Environmental change, ecosystem science, attitudes and values

Steve Trudgill

rentice
Hall

cf PEARSON EDUCATION

land • London • New York • Reading, Massachusetts • San Francisco • Toronto • Don Mills, Ontario • Sydney •
pore • Hong Kong • Seoul • Taipei • Cape Town • Madrid • Mexico City • Amsterdam • Munich • Paris • Milan

Pearson Education Ltd
Edinburgh Gate
Harlow
Essex CM20 2JE
England

and Associated Companies throughout the world

Visit us on the World Wide Web at:
http://www.pearsoneduc.com

First edition 2001

ISBN 0582-30347-8

British Library Cataloguing-in-Publication Data
A catalogue record for this book can be obtained from the British Library

10 9 8 7 6 5 4 3 2 1
05 04 03 02 01

Typeset in Garamond 11/12pt by 63
Printed and bound in Great Britain by T.J. International Ltd,
Padstow, Cornwall

DEDICATION

To "Uncle" Henry
who as a friend of the family left us so much literature
and a wealth of natural history and other books
that so much inspired my young imagination:
thanks to him, and my parents, I thought it was
normal to grow up surrounded by such literary riches.

and

To Dick Chorley
for telling me the apposite tale about the man with a beautiful garden
and someone said "you and God have done a good job"
and the reply came "yes, but you should have seen it when God had it to himself"
... and also
who on learning that I had been given a room in College
soon after I arrived at Cambridge
said "Good. Now you can go and write books and *think*
– you don't have to have a grant to think",
thereby giving me the possibility of believing in Universities again.

Contents

Foreword

Stephen Trudgill
Department of Geography, University of Cambridge

The prospect of future environmental change challenges our science and our values. However, in the history of the earth, it is clear that change is normal, and humans and other organisms can be very adept at responding to change, but the levels, nature and timings of changes are now uncertain – and the rates of change may well be increasing. Many scenarios of future climatic change exist but often these are in many ways no more than extrapolations of what we already think we know. This book tries to pose the questions which underlie the many-sided debate of how to respond to and influence change: *How should we view nature? What do we do for the best – how should we act – what are we trying to achieve and what should we be guided by?* In doing so the book introduces and attempts to analyze not only scientific aspects of the debate but also cultural attitudes and values: the notions of ecosystem stability are now challenged and it is also clear that *ecosystems are renewable but not repeatable.* It finds that prescriptive "solutions" based on current constructs may not be adequate. Feeling that analysis should lead to advocacy, the author believes *that if we can't improve predictability, we have to increase adaptability* which means that ecological and social capacity building should be advocated. This is seen in terms of concepts, institutions, attitudes and values which allow for a plurality of meanings and which can cope with surprise and unforeseen change – and which also facilitates responses to change. This endeavour is actually seen as worthwhile independently and irrespective of any prospect of climatic change.

Stephen Trudgill lectures in Biogeography and Environmental Management at the Department of Geography at the University of Cambridge and is Geography Fellow at Robinson College, Cambridge where he is also Chair of the Gardens Committee. He is a member of the Executive Committee and the Science and Education sub-Committee of the Field Studies Council and Chair of its Working Party on Environmental Policy. He is a member of the Editorial Boards *of Geography Review* and *Ethics, Place and Environment* and is currently Chair of the Environmental Research Group of the Royal Geographical Society with the Institute of British Geographers.

Understanding global environmental change: themes in physical geography

Around the season of All Hallows' Eve, 1994, Tom Spencer invited Olav Slaymaker to embark with him on the ambitious project of developing a book series on physical geography designed to come to grips, at an undergraduate student level, with the ongoing revolution in environmental geoscience. This revolution is the third phase in a sequence of intellectual revolutions initiated by global plate tectonics, reinforced by profound new insights into the complexity of the worldwide events of the late Quaternary and now focusing around the theme of global environmental change. Whether one believes in the influence of Hallowe'en over the lives of people or not, the challenge was accepted and we have learned that, not only is the project timely for physical geography as we understand it, but that many other geographers and geoscientists of various stripes and persuasions have encouraged us to proceed.

The series is intended for use by first- and second-year level geography and environmental science students in universities or colleges of higher education in the UK. It is also suitable as text material for second- or third-year level geography and environmental science students in Canada and the USA. Our main assumption is that the reader will have completed an introductory physical geography course.

Our motivation is discussed more fully in Chapters 1 and 8 of Volume 1, but for both of us there is an urgency in launching this series that derives at one level from a concern with re-establishing the historical legitimacy of the physical geographic tradition within the academy. We perceive, rightly or wrongly, a crisis of confidence among geographers over the traditional strength of the field in establishing linkages between the natural and social science traditions; its achievements in regional and global analysis and its preoccupation with the Earth as the home of people over Holocene and contemporary timescales. At a deeper level, and in common with many thoughtful persons of goodwill, we are still impressed by the recklessness with which the natural spheres of Earth continue to be exploited and believe passionately that we are accountable to subsequent generations for the way in which we respond to both biogeochemically and societally induced global environmental change.

The first volume in this series engages the most general trends and concepts associated with global environmental change and traces some of their connections with classical physical geographic writings. Subsequent volumes look in greater depth at each of the Earth's spheres and their interactions. We are grateful to the outstanding scientists who have committed their expertise and we acknowledge the initial work of Sally Wilkinson, then Publisher with the Longman Group, in enabling this project's commissioning. Subsequently, the sterling work of Tina Cadle (Development Editor, Sciences) and of Matthew Smith (Commissioning Editor) with Pearson Education, have commanded our respect and appreciation. We are reminded that All Saints' Day normally follows All Hallows' Eve at the stroke of midnight. Though this series will take a little longer to complete, it is our hope that, in its final form, it will make a small but perceptible contribution to the understanding that students of the next generation will have of the Earth. While there is no shortage of concern over the environment, there is an enormous challenge in converting this concern into meaningful action. We hope that these volumes will show that the physical geographer does have an important and distinctive role to play in the global environmental change debate. We are all stewards of the Earth; if through education we can ensure that the next generation's contribution to global sustainability will be more effective than that of our own then we will have achieved an important aim.

August 1, 1997
Olav Slaymaker, Vancouver
Tom Spencer, Cambridge

Volume preface

When I began writing this book, the intention was that it was going to be a textbook based in many ways on my earlier *Soil and Vegetation Systems* book. Here, I had produced a number of diagrams which showed functional linkages in ecosystems; the simple proposition was to write about environmental change in terms of how changed conditions would alter the nature of these functional linkages. This is still a viable proposition in that the diagrams could be used to follow through the consequences of changes in, say, temperature and rainfall leading to changes in weathering, nutrient inputs and leaching. However, it soon became apparent that, firstly, this represented a particular view of ecosystems which could be, and was, contested. Thinking about responses to environmental changes led me to writings which challenged the nature and validity of our ecosystem models. I sympathized with a student who said that she used to know what to think at "A" level but now wasn't sure – which after all is simply the beginning of thinking in new ways. Secondly, it became clear that thinking about environmental change also challenged our attitudes, values and assumptions about nature. Cultural constructs of nature became as an important issue as the validity of the models and so man–nature relationships became a central theme.

Views of nature have changed from a mechanistic, compartmentalized complexity, which nonetheless might be controlled, to involve more fluid, open concepts of self-organization and mutual reactivity, with an attendant less predictable response. Enabling nature to respond to changed conditions became a key concept. In parallel with this, social themes emerged about access to ecosystem products and societal goals where the tempting "win-win" scenario of simultaneously enabling nature and society emerged. What follows therefore is not so much a textbook but a series of essays which aim to explore the consequences of environmental change for ecosystem science and also for our cultural attitudes and values.

Even though, as an undergraduate at Bristol I read Geography with Geology and Sociology (and I have to admit here that I only gained a bare pass in the latter), it makes me somewhat nervous in venturing into sociological fields. As I am writing I can hear both ecosystem scientists and social scientists tut-tutting at what I am sure they will find an inadequate representation of each of their fields. It has been a bit of a struggle, though also enjoyable, challenging and stimulating, for me to move from limestone geomorphology through soil and vegetation systems and environmental issues to then encompass both social and psychological issues. However, I am convinced about the need for an approach which encompasses physical science on the one hand and the social science approaches which involve examining concepts, attitudes and values on the other. Believing it to be important to do this, I thought "why not have a go myself?" This still seems reasonable, spurred on by the dual thoughts that I was confident that what I was writing was good and that I would succeed, and simultaneously the gnawing thought that it might not be and that I wouldn't. Both sets of thoughts are the necessary fulcrum of creativity (despite the fact that the latter sense of inadequacy has led to huge bills from Heffers Bookshop). At the very least it meant that I no longer kept all my "science" books in the Department and the literature, "cultural" and philosophical books at home, but brought them all together in college. As I write, Henry Williamson's *Lone Swallows*, Bill Bryson's *Walk in the Woods*, Thoreau's writings, Aldous Huxley's *The Human Situation*, Dunbar's *Trouble with Science*, Steinbeck's *Grapes of Wrath*, Bell *et al.*'s *Environmental Psychology*, Russell's *Soil Conditions and Plant Growth*, Bouwman's *Soils and the Greenhouse Effect*, Richards *et al.*'s *Geomorphology and Soils*, Retallack's *Soils of the Past* and Golley's *Ecosystem Concept in Ecology* all lie together in a confused but hugely enjoyable and stimulating jumble of sources on my desk. The unattributed poems in the text are, I am afraid, mine.

In undertaking this endeavour I might also record that I am no fan of C.P. Snow's famous two cultures, a concept which I believe presents a false dichotomy – both science and the arts are about deriving stories which make sense of what we experience. I am also encouraged by the number of writings which also tackle this necessary but challenging approach. For example, texts such as Berkes and Folke on building ecological and social resilience, Atkins *et al.*, on people, land and time, Berkes on traditional knowledge and resource management, Ghimire and Pimbert on social change and conservation, Perlman and Adelson on exploring values in conservation, to mention a few, all sit squarely in the key areas of the arena of the debates about the crucial questions: *How should we view nature? What do we do for the best – how should we act – what are we trying to achieve and what should we be guided by?*

In order to answer these questions we can no longer rely on answers from the sciences alone or, alternatively, say, on responses from social surveys on attitudes – both are important, and the dialogue between the two is crucial. I don't pretend to have all, or necessarily any, of the answers but the hope is to assist in the evolution of a next generation of students which is better equipped to face the challenges. Thus I have attempted to rehearse some of the issues involved in the belief that geographers, with their physical and human traditions, should be central to the theme of the future of man–nature relationships. Many authors discuss whether man is part of nature or separate from it – I see this as a rather pointless and sterile discussion. The real thing to say is that we are *involved* with nature, both conceptually and physically, even those people who live in cities. It therefore seems to me to be appropriate that we should discuss our involvement with nature and the nature of that involvement, which is what I have attempted to do. Reviewers will no doubt judge the wisdom of this attempt. At least, I do not need to reiterate the quote from *Puckoon* that the book nearly drove Spike Milligan mad, which I echoed in *Soil and Vegetation Systems*. In fact, having tried to make sense of many different aspects and points of view in the literature I can cheerfully say that this book nearly drove me sane!

Acknowledgements

I would like to thank the University of Cambridge Department of Geography for allowing, and indeed encouraging, me to pursue my varied interests and especially Keith Richards who as Head of Department has maintained an academic and intellectually led atmosphere in the department, thus enabling the diversity of approaches which has emerged in this book as well as enabling an environment where everything you propose is immediately challenged (Ah, But! What about …). In addition I thank Robinson College for providing a "safe" room for me in which I could write in a tranquil environment which is about as agenda-free as you can get and where I can preserve a world of my own making. Also the College Fellows and members of the Department of Geography who have wittingly or unwittingly contributed to the nature of this book – especially the Biogeography and Environmental Management undergraduates who have suffered my endeavours to develop my ideas in lectures and contributed in no small way to them through their pertinent contributions in supervisions and essays.

It has taken not a little courage to attempt to range from Steinbeck to chemical equations but my belief that cultural constructs, attitudes and values are as important as scientific understanding has only been bolstered by discussions with colleagues, postgraduates and undergraduates in the department and other Fellows and my students in college together with other friends. I would especially like to thank Jane Robinson for her library hunts and Tim Bayliss-Smith, Bill Adams, Keith Richards, Stuart Lane, Robin Glasscock, Phil Howell, Sue Owens, Gerry Kearns, Jim Duncan, Alan Baker, Satish Kumar, Laura Cameron, Peter Jackson, André Roy, Alistair Kirkbride (who, on having sight of an earlier version of what I was carefully trying to write simply said "why don't you write what you want to?") together with John Parker, Director of Cambridge University Botanic Gardens and particularly Angus Jeffries, (our joint postgraduate on climatic change and gardens) for reading through the manuscript and making invaluable comments. I must apologize to my other postgraduate, Richard McDowell who came to work with me on soil chemistry and found that not only had I moved on from that topic but also that when I was not teaching, I was always busy working on this book (but he seems to have coped very well without me).

There are many others I would like to thank including those who might have only made a casual comment over coffee or in seminars but have certainly set me thinking. In particular, I would like to mention members of the Field Studies Council, especially Sue Townsend, Tony Thomas and Tim Burt. In addition, Rita Gardner, Alison Glazebrook, Tim Unwin, Sam Berry, Sarah Parkin and other participants for assisting with the recent RGS Environmental Ethics and Education Conference and Ron Johnston who once exhorted me "to go and find out what human geographers actually did" and Ian Simmons who in no small way encouraged me in his review of my *Barriers to a Better Environment* book with the assessment that I had made the transition from physical to social science with a "seamlessness which many of us would envy" – this has kept me going more than once, especially as I have been very mindful of David Pepper's review of the same book that "the reader might just as well drink a cup of herb tea as to read this". Talking of tea, I should also thank Jimmy and Di McPetrie whose farmhouse kitchen in Slapton, Devon, has seen many a lively debate which always challenged every idea I have had, together with staff at Slapton Ley Field Centre, especially Keith Chell and Nigel Coles. Other people who have kept me going have included Elizabeth Burt who suggested I frame my watercolours and made me believe that they weren't that bad and, of course, include "the village people" of Hinxton like Pip Arran and especially, to name but a few, Jill, Simon, Wendy and John who provide cups of coffee and the latter two who also provide endless entertainment through their garden alterations (this week it's the new pond) all of which keeps my feet on the ground. I can't thank Robinson College Gardens Committee enough for making me realize that it can be impossible to find answers – as Chair of the Committee, I often listen to impossibly implacably opposed differences of opinion and have learnt the phrase from the Junior Bursar: "There seems to be no measure of agreement here". A famous incident, quoted by the Senior Tutor, occurred once when two motions were tabled, one referring to the "glorious patch of wild flowers", the other to the "area of noxious weeds" which, as you might guess, both referred to the same area. This teaches us all something about cultural constructs of nature.

Talking of people seeing things differently, when I was younger, I readily identified with Gerald Durrell's *My Family and Other Animals*: Gerald busily grubbed in the soil, entranced by natural history, while his elder brother Laurence had his head stuck in a book. This almost exactly mirrored how me and my brother were. Perhaps now that I might have seemed to have escaped from my purely natural history mould, I wonder if my brother will extend his interests from linguistics to gardening? Perhaps not – people *are* different – and that has also taught me something about allowing for people's diversity rather than being vainly prescriptive about what people should be interested in!

I should also thank Tom Spencer for comments on the manuscript, Sally Wilkinson, formerly of Longman, and Matthew Smith of Pearson Education who has been so patient in awaiting the manuscript.

Finally I would like to thank my students in college, who after a discussion of whether or not soil was a renewable resource, kept me amused by giving me a bag of potting compost, saying "of course it's renewable, you can buy it at a shop!" I will refrain from a comment about attitudes and values among young people today.

Stephen Trudgill
Robinson College
23 February 2000

The publishers wish to thank the following for permission to reproduce the material: Invocation from Seed, J., Macy, J., Fleming, P. and Noess, A. (1993) *Thinking like a mountain, towards a council of all beings*, with the permission of New Society Publishers. Figure 1.1(a) adapted from Adams, J, Maslin, M and Thomas, E. (1999) 'Sudden climate transitions during the Quaternary', *Progress in Physical Geography*, 23, 1, 1–36, Arnold; Figure 1.1 (b) from Leaky, R & Lewin, R (1995) *The Sixth Extinction: Biodiversity and its survival* by permission of Sherma BV; Figure 1.3 reprinted from *Weathering, Soil and Paleosols: Developments in Earth Surface Processes*, vol. 2, Retallack, G.J. 'Paleozoic paleosols', 543–564 (1992) with permission from Elsevier Science; Figure 1.4 from Costanza, R. *et al*, 'The value of the world's ecosystem services and natural capital', *Nature*, 387, 15 May 1997, ©(1997) Macmillan Magazines Limited and by permission of Robert Costanza; Figure 1.5 (a, b and c) from Matthews, E., (1984) Global inventory of pre-agricultural and present biomass. *Progress in Biometeorology*, 3, 237–246, Backhuys Publishers b.v. Figure 1.6 George, M (1997) *The Norfolk Broad*, Transactions Norfolk & Norwich Naturalists' Society 24 [2] by permission of the Norfolk and Norwich Naturalists' Society and Dr. M. George; Figures 1.7 and 1.9 from Clark, William C. (1989) 'The human ecology of global change', *International Social Science Journal*, 121 *Reconciling the Sociosphere and the Biosphere* 315–345, Blackwell Publishers; Figure 1.8 from Barry, R.G. & Chorley, R.J. (1998) *Atmosphere, Weather and Climate* and Figure 1.11 from Bennett, R.J. & Chorley, R.J. (1978), *Environmental Systems*, (Fig. 9.1, page 469) by permission of Routledge Publishers; Figure 2.1 from Bell, P.A., Greene, T.C. Fisher, J.D. and Baum, A. (1996) *Environmental Psychology*, Academic Press Limited; Figure 4.1 reprinted from Mitchell, T.D. & Hulme, M. (1999) Predicting regional climate change: living with uncertainty. *Progress in Physical Geography* 23, 1, 57–78, p.67, Arnold; Figure 4.2 reprinted from *Journal of Hydrology*, vol 205, Burt. T.P., and Shangedanova, M. 'An historical record of evaporation losses since 1815 calculated using long-term observations from the Radcliffe Meteorological Station, Oxford, England', pages 101–111 (1998) with permission from Elsevier Science; Figure 4.4 from Houghton, J. *Global Warming: The complete Briefing*, 2nd edition(1997), Cambridge University Press; Figure 4.5 from Godwin, H. *A History of the British Flora* (1975), Cambridge University Press; Figures 4.7 (a) from Huntley, B. Berry, P.M., Cramer, W. and McDonald, A.P. (1995) 'Modelling present and potential future ranges of some European higher plants using climate response surfaces', *Journal of Biogeography*, 22, 967–1001 (Fig. 15) by permission of Blackwell Science; Figure 4.7 (b) and (c) from Jalas J. & Suominen J. (1976) *Atlas Florae Europaeae: Vol. 2 Gymnospermae (pinaceae to ephedraceae and Vol. 3 Salicaceae to Balanophoraceae* by permission of Societas Biologia Fennica Vanamo, Botanical Museum, Helsinki; Figure 5.2 from Friday, L. and Rowell, T. (eds), (1997) *Wicken Fen: The making of a wetland nature reserve* (Fig. 62, p.211), by permission of Laurie Friday, Terry Rowell (original author of the Figure) and Harley Books; Figure 5.9 from Jones, M. and Talbot, E. (1995) 'Coppicing in urban woodlands', *Journal of Practical Ecology and Conservation*, 1, 1, 46–52 (Fig. 1, p.47) by permission of Wildtrack Publishing and the authors; Figure 5.12 from Porteous, J.D. (1996) *Environmental Aesthetics: ideas, politics and planning* (Fig. 3.16, p.126) by permission of Routledge Publishers; Figure 6.3(a) from Englesad, O.P., Stroder, W.D and Dumenil, L.C. (1961) 'The effect of surface soil thickness on corn yields', *Soil Science of America Proceedings*, 494–499 (Figures 2 and 3), The Soil Science Society of America; Figure 6.4 from Shrader-Frechette, K. and McCoy, E. *Methods in Ecology: Strategies for conservation* (1963), Cambridge University Press;

Figures 6.6 (a, b and c) reproduced from Soil Survey and Land Research Centre, Matthews, B. (1971) *Soils in Yorkshire I (Sheet SE65, York East)*, Soil Survey Record 6, © Cranfield University, 1971. No part of this publication may be reproduced without the express written permission of Cranfield University; Figure 7.1 from Trudgill, S. 1989, 'Soil types: a field identification guide', *Field Studies* 7 (2), pp. 337–363 by permission of The Field Studies Council; Figure 7.5 from Retallack, G.J. (1996) 'Palaeosols: Record and engine of past global change', pp. 25–28, reprinted with permission from *Geotimes*, June 1996, © 1996, American Geological Institute; Figure 7.10 from Ellis, S. and Mellor, A. (1995) *Soils and Environment* (Fig. 3.1, p.59) by permission of Routledge Publishers; Figure 7.11 from Huang, W.H. and Kiang, W.C. (1972) 'Laboratory Dissolution of plagioclase feldspars in water and organic acids at room temperature', *American Mineralogist* 57, 1849–1859 (Fig. 4), American Mineralogist Society of America; Figure 7.12 from Trudgill, S.T. (1985), *Limestone Geomorphology* (Longman) by permission of Pearson Education Limited; Figures 7.14 (a, b and c) from Haynes, R.J. (1986) *Mineral Nitrogen in the Plant-Soil System*, Academic Press Limited; Figure 7.18 from Ineson, P., Taylor, K., Harrison, A.F., Poskitt, J., Benham, D.G, Tipping, E. and Woof, C. (1998) 'Effects of climatic change on nitrogen dynamics in upland soils. 1. A transplant approach'. *Global Change Biology*, 4, 143–152 (Fig. 2, page 145) by permission of Blackwell Science; Figure 7.19 from Thornley, J.H.M. and Cannell, M.G.R. (1997) 'Temperate grassland responses to climate change: an analysis using the Hurley Pasture Model', *Annals of Botany*, 80, 205–221, Academic Press Limited; Table 8.1 from Hannah, L., Hutchinson, C., Carr, J.L. and Lankerani, A., (1994) 'A preliminary inventory of human disturbance of world ecosystems', *Ambio*, 23, 4-5, 246–250; Figure 8.5 from Bowler, I. *Agricultural Change in Developed Countries*. (1996), Cambridge University Press; Table 8.6 from Bennett, A.J. (1994) 'Soil science and better land use in the tropics' Chapter 20 in Syers, J.K. and Rimmer, D.L., *Soil Science and Sustainable Land Management in the Tropics*, CAB International; Figure 8.8 from Briggs, D.J. & Courtney, F.M. (1985) *Agriculture and Environment* by permission of Pearson Education Ltd; Figure 8.9 reprinted from *Global Environmental Change, vol. 9*, Brookfield, H. and Stocking, M., 'Agrodiversity: definition, description and design', pages 77–80 (1999) with permission from Elsevier Science; Figure 8.10 reproduced from Institute of Hydrology Annual Report 1995–6 (Dr. N.A. Jackson), by permission of Centre for Ecology and Hydrology, Wallingford. Figure 9.3 from Sack, R.D. 'The realm of meaning: the inadequacy of human-nature theory and the view of mass consumption', Chapter 40 in Turner, II, B.L., Clark, W.C., Kates, R.W., Richards, J.F., Matthews, J.T. and Meyer, W.B. (1993) *The Earth as Transformed by Human Action: Global and regional changes in the biosphere over the past 300 years*, 659–671 (1990); Figures 9.4 and 9.5 from Agarwal, A. and Narain, S. (1990) *Towards green villages: a strategy for environmentally sound and participatory development*, New Delhi, Centre for Science and Environment by permission of Amil Agarwal and Sunita Narain; Figure 9.6 reprinted from *Global Environmental Change*, vol. 9, O'Riordan, T. and Jordan, A., pages 81–93 (1999) with permission from Elsevier Science. Plate 1: Photo accompanying 'Lost Wildlife . . .' © Martin Pope; Photo accompanying 'Welcome to . . .' The Independent Syndication/Andrew Hasson; Headlines/articles: 'Decay' (Anna Pavord) 10 June 1999, 'Not-so-noble' (Mark Rowe) 30 May 1999, 'They're going under' (Geoffrey Lean) 13 June 1999, 'Brazil moves' (Geoffrey Lean) 4 August 1996, 'Welcome to' (Marina Baker) 8 July 1999, 'What to do' (Yvonne Cook) 1 June 1999, 'Lost Wildlife' (Michael McCarthy) 8 July 1999, The Independent Syndication.

Though every effort has been made to contact the owners of copyright material, in a few cases this has not been possible and we take this opportunity to apologise to any copyright holders whose rights may have been unwittingly infringed.

Invocation

Invocation, from a Deep Ecology book called *Thinking Like a Mountain, Towards a Council of All Beings* by John Seed, Joanna Macy, Pat Fleming and Arne Naess, paperback ISBN: 0–86571–133–X. The invocation was written by John Seed.

We ask for the presence of the spirit and pray that the breath of life continues to caress this planet home.

May we grow into true understanding – a deep understanding that inspires us to protect the tree on which we bloom, and the water, soil and atmosphere without which we have no existence.

May we turn inwards and stumble upon our true roots in the intertwining biology of this exquisite planet. May nourishment and power pulse through these roots, and fierce determination to continue the billion-year dance.

May love well up and burst forth from our hearts.

May there be a new dispensation of pure and powerful consciousness and the charter to witness and facilitate the healing of the tattered biosphere.

We ask for the presence of the spirit to be with us here. To reveal to us all that we need to see, for our own highest good and for the highest good of all.

We call upon the spirit of evolution, the miraculous force that inspires rocks and dust to weave themselves into biology. You have stood by us for millions and billions of years – do not forsake us now. Empower us and awaken in us pure and dazzling creativity. You that can turn scales into feathers, seawater to blood, caterpillars to butterflies, metamorphose our species, awaken in us the powers that we need to survive and evolve into more aeons of our solar journey.

Awaken in us a sense of who we truly are: tiny ephemeral blossoms on the Tree of Life. Make the purpose and destiny of that tree our own purpose and destiny.

Fill each of us with love for our true Self, which includes all of the creatures and plants and landscapes of the world. Fill us with a powerful urge for the well-being and continual unfolding of this Self.

May we speak in all human councils on behalf of the animals and plants and landscapes of the Earth.

May we shine with a pure inner passion that will spread rapidly through these times.

May we all awaken to our true and only nature – none other than the nature of this living Earth.

We call upon the power which sustains the planets in their orbits, that wheels our Milky Way in its 200-million-year spiral, to imbue our personalities and our relationships with harmony and joy. Fill us with a sense of immense time so that our brief, flickering lives may truly reflect the work of vast ages past and also the millions of years of evolution whose potential lies in our trembling hands.

O stars, lend us your burning passion.

O silence, give weight to our voice.

Part I

Concepts, attitudes and values

Chapter 1

Introduction: ecology for people

Summary of key points:

1. Environmental change challenges ecosystem science and our fundamental, underlying attitudes and values.
2. If we can't improve predictability, we have to improve adaptability.
3. "Knowing the 'why' is essential if we expect to change 'what' we do" (Sack, 1990).
4. The fact of dependency on ecosystems does not vary, the degree of directness of the dependency and access to resources do.
5. "The Greens" have often used an evangelical zeal which doesn't work as it alienates many people; it would be better to try for a "win-win" where ecology is *for* people in a combined ecological, economic and social justice context.

1.1 Ecosystems, change, ecology and people

... everyone spoke of liberty, social justice and democracy, but liberals, catholics and communists all had their own definitions ... and the promotion of [their] ideologies. What politics failed to do was to meet people's perceptions of their own needs – security, shelter, fuel, family, food and livelihood.

Kinta Beevoir (1993) *A Tuscan Childhood*, Penguin Books

... the trouble with you in the west   You don't really value human life, human beings. You have to learn ... that a human life, any human life is worth more than a car, or a plant, or a tree ... or a monkey.

William Boyd (1991) *Brazzaville Beach*, Penguin Books

The fundamental task for environmentalists is not just to save nature. They need to find and promote answers to the moral questions ... environmentalists should leave most of their ecological ammunition at home and instead encourage people to address the more fundamental issues. What kind of world do we want to live in? What kind of social and natural landscape would describe this world? What do we do, individually and collectively, to help our little part of the world in this direction? No one should expect this discussion to lead to easy consensus ... but this task of re-thinking ethics will surely result in a more inclusive environmentalism, one uniting it with other social movements in a common moral cause; to help create a more liveable world for us all, humans and nonhumans alike

J.D. Proctor (1995) Whose Nature? The contested moral terrain of ancient forests. In W. Cronon (ed.) (1995a) *Uncommon Ground: Toward reinventing nature*, W. Norton

It is hard to escape from the context of global warming today, but I believe there are more fundamental issues at stake. The thing about global warming is not to just consider whether you believe in it or not, nor how valid the predictions of changes and impacts are, though these topics are the focus of much discussion. More fundamental are the underlying questions about man–nature relationships and our attitudes and values to and about the environment and environmental science. The good thing about global warming is that it has thrown these issues into sharper focus and, to my mind, put them centre stage. The studies of global warming, and environmental change in general, are legitimate topics for study; however, there is the deeper study of the bases of attitudes and concepts which underpin all our interactions with the environment: as Vitousek (1994) has said, there is a study which is "beyond global warming".

If you look back through the millions of years of geological history of the earth (Table 1.1) it is clear that climatic change is normal. However, currently, in a study of climatic change, Mitchell and Hulme (1999) are clear that the task of scientists is still to counter the damaging norm that we now have the prospect of departing from a stationary climate. If we look back, such a norm is clearly untenable: warming, desertification, cooling, and ice ages have all occurred in turn and left their traces in the geological record. These changes are partly because the continents have moved through different climatic zones during continental drift but also in terms of shifts in global climate. Throughout these changes, land covers (forests, grassland or ice caps, for example) have shifted and

Table 1.1 Changes in the geological past. See also Figure 1.1(b)

Era	Start date (million years)*	
Holocene		
Pleistocene	2.4–3	Mankind appears
Tertiary	65–70	Flowering plants in full development, great variety of mammals
		Mass extinctions
Cretaceous	135	Dinosaurs and ammonites extinct
Jurassic	180–205	Birds and mammals appear
		Mass extinctions
Triassic	225–250	Flying reptiles and dinosaurs appear
		Mass extinctions
Permian	270–290	Rise of reptiles and amphibians
Carboniferous	350–355	Coal forests
		Extinctions
Devonian	400–410	
Silurian	440–435	
		Extinctions
Ordovician	500–510	
Cambrian	570–600	
Precambrian		

* Different texts give varying dates

many species have become extinct and others have evolved. Even in the recent past, and to the present day, climate has shifted considerably on the scale of decades (Anderson and Willebrand, 1996); Adams *et al.*, (1999) write of sudden climate transitions in the last few hundred thousand years (Figure 1.1) and, indeed, Sloan and Rea (1996) also point out that the Eocene period (see Table 1.1) was the "warmest time interval of the past 65 million years".

What is different now is that we are dealing with human values and attitudes: perceptions, preferences and perhaps guilt and fears. In addition, we see the changes as having trends, and indeed consequences, which are not necessarily predictable: Mitchell and Hulme (1999) in a treatise on living with uncertainty write that: "regional climate prediction is not an insoluble problem, but it is characterized by inherent uncertainty". We also now perceive that current environmental changes may be related to our activities and we regret losses of species and vegetation types, fearing the consequences for the future and expressing preferences for states of the world. These preferences are in terms of the value of a state to us, in both a spiritual and a utilitarian sense, with the perception that the changes may threaten the ecosystems on which we depend. Despite the fact that there have been many changed states in the geological past, the preferences also relate to the "guilt-laden" notion that human-induced changes have somehow damaged a "pristine" nature. Glacken (1967) writing in *Traces on the Rhodian Shore* feels that there has been a persistent question in Western thought which was crystallized in Marsh's (1864) *Man and Nature*: "In his long tenure of the earth, in what manner has man changed it from its hypothetical pristine state?" The ideas of "balance" or "equilibrium" and the corresponding human "disturbance" have become firmly entrenched in Western culture. There are thus fears for the future of nature and fears for the continued existence of mankind. It may be a human conceit, but we also see the changes we have made as more extensive and more far-reaching and threatening than any that have preceded. However, by contrast, some writers, for example Robinson *et al.* (1998), welcome the increased levels of carbon dioxide, refuting the global warming hypothesis (as discussed in Chapter 4) and focusing on the increased plant growth which occurs at enriched CO_2 levels. They propose that there will be a great increase in plant – and agricultural crop – growth, thus benefiting mankind by providing an increasingly lush environment of plants and animals and more biotic resources. Warrick *et al.* (1986), writing on the effect of increased CO_2 and climatic change on ecosystems felt that "there are no firm grounds for believing that the net effects of increased CO_2 and climatic change will be adverse rather than beneficial". In terms of predictions, early attempts at the predictive modelling of climatic change were critically appraised by Schlesinger (1991) where the conclusion was that at scales less than the continental, the predictions were poor, though this situation has improved in recent years (see Houghton's review, 1997).

Public perception might range from indifference to alarm and there is a certain "hype" to media coverage (see Plate 1.1). There are books like Erickson (1990) which talk of "Greenhouse Earth: Tomorrow's disaster today" and research papers such as Epstein (1995) on "Emerging diseases and ecosystem

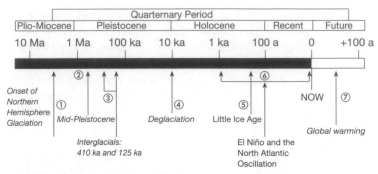

Note that the timescale is logarithmic:

(1) Between 4 and 2.5 Ma ice sheets started to develop in the Northern Hemisphere.
(2) Prior to the mid-Pleistocene the climate cycled between glacial–interglacial every 41 ka; afterwards it cycled every 100 ka.
(3) The two closest analogues to the present climate are the interglacial periods at 420–390 ka and 130–115 ka (also known as the Eemian).
(4) Deglaciation.
(5) Little Ice Age (AD 1700).
(6) El Niño (~3–5 years) and North Atlantic climate oscillation (~10 years), which have occurred for at least the last 1000 years.
(7) Anthropogenic global warming.

(a)

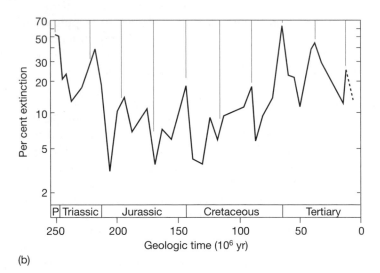

(b)

Figure 1.1 (a) Climatic change in the recent past from the later Tertiary about 10 million years (Ma) ago, the Plio-Miocene, through the Pleistocene which started about 1 Ma ago, the post-glacial (Holocene) and Recent periods, the last three making up the Quaternary period (modified from Adams *et al.*, 1999). (b) There have been many extinctions in the geological past. Despite this many people feel that it is not right for people to be responsible for extinctions, arguing also that the current rate of extinctions in relation to human activity is much more rapid than during the geological past (modified from Leakey and Lewin, 1995).

Plate 1.1 Dramatic headlines involving the environment and global warming appear almost daily. Credits/sources see acknowledgements.

instability – new threats to public health" writing of a "global resurgence of infectious diseases as ... stresses cascade through the community assemblage of species" and concerns expressed about investment (Forsyth, 1999). There are scientific uncertainties such as are encapsulated by the title of the paper by Carlson and Bunce (1996): "Will a doubling of carbon dioxide concentration lead to an increase or decrease in water consumption by crops?" (a small seasonal increase in transpiration, and thus water use, is, in fact, indicated in their study). There are also many reviews, for example Sulzman *et al.*'s (1995) "Modelling Human Induced Climatic Change – A Summary for Environmental Managers" and reviews of the Kyoto Protocol (Grubb *et al.*, 1998) and the refreshingly sober summary on the greenhouse effect in Kemp's (1997) useful *Global Environmental Issues* book. Finally, Burroughs (1997) in *Does the weather really matter? The social implications of climatic change* has a rather bleak view: "When surrounded by seemingly impenetrable thickets of uncertainty, crashing off into the undergrowth in search of a way out is a risky option" and concludes that the public will be "told to cope with whatever is thrown at them" and ends with the statement that "just as I thought, *more research is necessary, but it won't give you the answers*" (emphasis added). The author does, however, make the plea for gradualism in progress and makes perhaps the point that while we might endeavour to plant trees for carbon uptake, isn't it a good idea to plant trees anyway?

Few things have thus posed more fundamental challenges to our understanding of the way the world works more than the predictions of climatic change, the discussions of the possible consequences and the significance of these consequences. Thus, our scientific approaches, social constructs, psychological attitudes, cultural values and spirituality are all challenged. As Irwin (1995) has noted, "*environmental threat does not sit apart from a wider range of threats to our sense of identity and security*" (emphasis added). There exist uncertainties in the science, contested interpretations, vested interests, social implications and challenges to fundamental assumptions. It is important, then, to take a broad approach to the topic of the terrestrial biosphere and environmental change and not only to look at global warming but also to consider our overall role in environmental change, our ecological and environmental knowledge, our needs and our responsibilities. These involve *examining both our scientific assumptions and our social values*. The propositions for tackling global warming range from burying carbon dioxide in the deep oceans (reported in *Chemistry and Industry*, 17 May, Anon. (1999)) to carbon sequestration by planting trees (reviewed recently by Henderson, 1999) to calls for changing views to enable "discourses (which) allow a critical evaluation of different versions of environmental issues" by Blaikie (1995), who also feels that appeals to more and better environmental science, better law and governance and to the logic of the market will only be partly successful. There is thus a wide range of responses. Indeed, to me this is the interest in global warming, not so much whether it will happen, or is happening, or not (we shall presumably find out as we go along) but the challenges and range of responses it has thrown up.

My approach is summed up neatly for me by Sack (1990), in that it: "involves understanding not only what we are doing to affect nature, but also *why we do it*. Describing 'what' in the sense of volumes of carbon dioxide released into the atmosphere, the amount of soil depleted … can be accomplished up to a point without considering social and individual motivations. These motivations, however, must be included in the 'why' of human behaviour, and *knowing the 'why' is essential if we expect to change 'what' we do"* (my emphasis). Why we behave the way we do means understanding our-selves as agents and the kind of life we wish to lead, and this understanding may well raise questions about ourselves that are difficult for conventional scientific methods to handle. Indeed, Davison (1998) goes on from this point to say that there *are* dynamical relations which are hard for classical science to handle and asks for a truly behavioural approach to the dynamic behaviour–environment system. Rather more graphically, Malone (1992) saw three possible options for the future – the end of the world; an increase in inequality, or *an attractive future through the use of creativity and imagination*. To achieve the third option a "fundamental re-orientation" involving "*the confluence of science, technology and society*" is needed. If this occurs, the author is then hopeful of "a marvellous, exciting and beautiful human prospect". These writings to me justify the broad range of approaches which I have adopted in the book which includes discussions of social attitudes and meaning as much as of scientific understanding.

If the role of mankind is now said to dominate the global current changes, then the assumption is that we are faced with a responsibility for the future as never before: an attitude that is encapsulated by Hayward (1994) who writes in his book, *Ecological Thought* that "*as our power to influence the destiny of the planet increases, so does our responsibility*" (emphasis added). This, however, leads onto the key question of: just how much influence do we have? In the study of the linkages between our activities and the state of the world, where do we stand in the spectrum from control to passivity in terms of earth processes? The answer will, of course, vary and the debate also involves discussing the evidence for causal linkages (and indeed the very nature of the concept of cause and effect) as well as the desirability and meaning of particular states of the environ-ment. Central is the debate about how the world reacts to change. This is particularly in terms of any adjustment or self-regulation which maintain cur-rent states and cumulative changes which can lead to new states, together with attendant judgements on the desirability of these states. There is a Gaian view that life creates conditions for itself on earth (Lovelock, 1995) which can be contrasted with the Erebian view (Retallack, 1992, after Ereban, the primeval god of darkness). Rettalack sees the question as not whether life has absolute control over the earth (Gaian) or no control (Erebian), but rather it is one of "how much?" His diagram (Figure 1.2) from a study of paleosol evolution (soils being seen as both a reflector and facilitator of life) shows that he adopts an intermediate view between the contested extremes.

There is also a need to re-examine cherished concepts such as resilience, disturbance and recovery and the basis of the ecosystem concept such as were expressed for example, by Walker (1989) in "Diversity and Stability" and by Holdgate and Woodman (1978) in *The Breakdown and Restoration of Ecosystems*. While we appeared at one time to be heading for integration through "unifying concepts" (van Dobben and Lowe-McConnell (1975) in "composite sciences" (Osterkamp and Hupp, 1996) *many ecological concepts often refer to preferred states of nature rather than to necessity in the real world*. Simmons (1993a) in

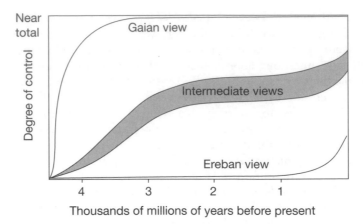

Figure 1.2 Views of the way life may have controlled the environment (Gaian view) or life may have been at the mercy of the environment (Erebian view) and what is viewed as a more reasonable view by Retallack (1992) from the evidence of traces of life found through his study of palaeosols (from Retallack, 1992).

Environmental History warned that we run into the danger of simply projecting onto nature "what the author wanted to believe anyway" and as Sheail (1995) writes there are also misgivings and debates within ecology as to the applicability of previously cherished concepts. Many concepts have thus been challenged by ecological scientists themselves with the concept of *continual dynamic change* now perhaps coming to the fore, as Pahl-Wostl (1995) concludes (which, of course, still may be what the author wants to project onto nature anyway). That author also feels that "Scientific knowledge does not simply emerge in the brains of ingenious researchers enlightened to reveal the truth in nature. Knowledge must be seen from a relational perspective depending on the values, beliefs, perceptions of nature and society." It would thus seem to be important to examine our ecological concepts of the world as social constructs of nature. This is because the ways in which we view the world not only dominate how we treat it but also because constructs are relative to our own needs which are material, cultural and spiritual. Central also is the relationship between ecological science and political ecology whereby we tend to propose plans for action and behaviour based to some degree on our understanding of the way the world works (Hayward, 1994). As Simmons (1993a) has already pointed out, this is a critical area of discussion because it becomes obviously flawed, and at least tautological, if we attempt to read moral and ethical tenets from the natural world when the natural world is a construct itself based on our values, beliefs and perceptions (Proctor, 1998a). In this context, Zimmerer (1996a) quotes Blaikie and Brookfield (1987) who saw political ecology as combining the concerns of ecology and a broadly defined political economy, encompassing a shifting dialectic between (and within) classes and groups within society and land-based resources. Zimmerer feels that the challenges which lie ahead in fact involve moving from the concept of the environment solely as a receptor of modification to one where environmental modification is seen as more fully and recursively integrated with theories of development.

Concepts are also challenged in another relational way. The real point about global environmental change is that change can be stimulating, and bring out the best of inventiveness in people; but it can also be psychologically disturbing in terms of bringing a sense of insecurity about relationships with the things around you and with other people. We all live by a set of assumptions and constructs about how the world works. Challenges to these assumptions and constructs brought about by change can thus make us uncertain about how we relate to the world and about the future. This can bring a sense of denial of the possibilities of change or a preference to ignore the changes even, and in some ways, especially, if the changes are framed in the form of dire warnings about catastrophe and losses. In an uncertain world, people inevitably cling onto notions of stability which is why notions of disturbance followed by recovery in ecosystems have their psychological appeal. However, even such cherished notions are often illusory because although a system might be perceived to have recovered from a disturbance, it is never the same as it was before because it has gone through a transition through time to get to the restored state and there will be legacies of that transition: *ecosystems are renewable but not repeatable.*

If it is accepted that ecosystems have always been changing through geological time and that species have come and gone then clearly continual change is, in fact, a dominant feature of ecosystems (Keulartz, 1998, p. 153). Rather than exhibiting disturbance and recovery to a stable state, Pahl-Wostl (1995) describes a "new concept of ecosystems as self-organising systems operating far from a stable equilibrium point". If this is so, what we then have to do is to examine our values and constructs about stability and change, to judge their appropriateness and adapt them to changing situations. Then, if new situations are inevitable, *and if we can't improve predictability, we have to improve our adaptability.* If the new situation is unacceptable we have to endeavour to change it to a preferred situation. Discussions of preferred states, inevitability, and the degree of influence we have on these states are thus further important themes. However, while we might accept and adapt to what we cannot change, changing what we cannot accept involves a concord on what we should do which in turn is based on predictions of the likely outcome of our actions. The latter is often not easy. Developing adaptability is thus going to be crucial because there are many possible reactions and the scenarios about the future which are essentially untestable until we get there.

A rationalist view of future states and preferences might be as:

FUTURE STATE	NOT PREFERRED	PREFERRED
Inevitable	Act to change it	Accept
Likely	Prepare to act	Prepare to accept
Unlikely	Accept the risk	Act to cause

However, future scenarios involve reliable prediction and preference involves consensus so the view that involves both uncertainty and a range of cultural responses might be more appropriate:

FUTURE	ACTION
Difficult to say	Prepare for combinations of eventualities through capacity building, technological development, flexibility and adaptability

Many writers are indeed pessimistic about our predictive abilities. This, per-haps, reflects upon the nature of our modelling approaches, which could be too mechanistic and compartmentalized rather than synergistic, alternatively they could be synergistic but the situation is too mutually interactive for us to unravel the complexities. In his book: *Interpreting Nature: Cultural constructions of the environment*, Simmons (1993b) writes that "Most science provides a … mechanical view of the world, with direct chains of cause and effect … The world is seen like a snooker table whereas the ecological view is that of a mobile in which all parts adjust to a change in the position of any one, though within constrained limits." (And if you are not sure what a mobile is – they were all the rage in the sixties – imagine that the snooker balls are variously attached to each other by lengths of string, or better still by lengths of elastic, so that when one ball moves, the rest also shift in a mutually responsive, but not very predictable, way.) He concludes that it is the complexity of this fluid interaction which means that "the long range consequences of human-induced alterations of ecological systems are poorly predictable". Yet, still there is the demand to know what the effect of a proposed course of action will be *as if* there were clear cause and effect linkages. We should therefore examine closely the nature of the con-structs used in our modelling of how the world works. A basic message is that it is certainly not the case that we have a fixed set of ecosystem states at the moment and we are trying to predict future changes from a fixed base line. It is the case that during a scenario of continual past and present change, we are try-ing to understand the nature of these changes and think how we might predict future changes, as well as how the changes might affect our dependencies on ecosystems.

The land surface of the earth covers some 153 230 000 km^2 (Costanza *et al.*, 1997). Of this, urban areas constitute 2% and cropland some 9% (Tables 1.2, 1.3, Figures 1.3, 1.4). Of the population of the earth, some 50% lived in urban areas in 1991 (Douglas, 1994). While these figures might be contested in terms of the definitions of land cover types (see Meyer and Turner (1994) for dis-cussions of definitions of each land cover type), it is clear that much of the world's population lives in a very small area (43% of the world's population on 1% of the total land surface, Miller (1988) quoted from Meyer and Turner (1994) q.v.), often without any day-to-day intimate contact with the terrestrial biosphere. However, despite this lack of contact our population still depends

Table 1.2 Estimates vary for the world cover of vegetation types according to definitions of types and scales. Version of Costanza et al. (1997). Compare with Table 1.3

Vegetation type	Area km²	% terrestrial		km²		km²
Forest	48 550 000	31.7	of which:	Tropical: 19 000 000	(% terrestrial: 12.4) (% of forest: 39.1)	Temperate/boreal: 29 550 000 (% terrestrial: 19.3) (% of forest: 60.9)
Grassland	38 980 000	25.4				
Desert	19 250 000	12.6				
Ice and rock	16 400 000	10.7				
Cropland	14 000 000	9.1				
Tundra	7 430 000	4.8				
Wetlands, lakes and rivers	5 300 000	3.5				
Urban	3 320 000	2.2				
		% global				
Terrestrial total	153 230 000	29.7				
Marine	362 020 000	70.3				
Global total	516 250 000					

Table 1.3 Estimation of world vegetation cover, version of Matthews (1984, 1983). Compare with Table 1.2

Matthews (1984)	km²	% of list
Tropical rainforest	12 300 000	21.7
Other tropical evergreen forests	3 300 000	5.8
Savannah (10–40% tree cover)	6 500 000	11.5
Cool deciduous woodland	9 200 000	16.3
Grassland, no trees	7 700 000	13.6
Cultivated area	17 600 000	31.1
	56 000 000	100.0

Matthews (1983)	km²	%
Forest + woodland	5.24E+09	39.6
Grassland	2.74E+09	20.7
Cultivated	1.76E+09	13.3
Desert	1.56E+09	11.8
Shrubland	1.21E+09	9.1
Tundra	7.30E+08	5.5
Marsh/swamp	2.00E+08	1.5
Total	1.34E+10	

Forest type		
Tropical rainforest	1.23E+09	31.2
Tropical/sub-evergreen seasonal	3.30E+08	8.4
Subtropical rainforest	2.00E+07	0.5
Temperate/sub-polar rainforest	4.00E+07	1.0
Temperate evergreen seasonal	8.00E+07	2.0
Dry evergreen forest	5.00E+07	1.3
Evergreen needle subtropical forest	5.00E+07	1.3
Evergreen needle temp. forest	9.30E+08	23.6
Drought-deciduous forest	2.90E+08	7.4
Cold deciduous forest	9.20E+08	23.4

upon the consumption from the agricultural area of crops which in itself is a relatively small proportion of the earth's surface.

Environmentalists told us that the world is under threat from rising population and also Western consumption. Steve van Matre (1983) in *The Earth Speaks* uses "a spaceship earth" analogy: "To support our continually expanding population of human passengers we are systematically destroying or enslaving much of the other life of the ship we sail upon. And we have used the fossil sunlight and water ... to support ... a relatively small number of us ... in amazing luxury ... most of us in western society live in ... synthetic, unnatural settings like swollen grubs feeding off of the energy-rich micro-environments we have created, all the while spewing forth our resulting disorder and poisonous wastes to contaminate other areas of the vessel." It has been calculated that to provide

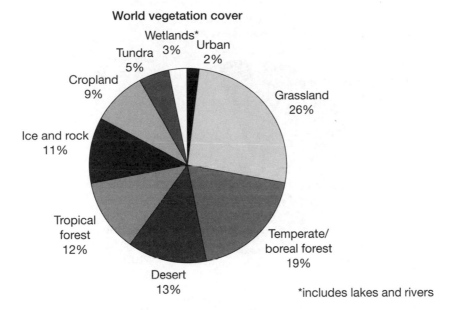

Figure 1.3 Estimates of world vegetation cover vary, according to definitions and scale. Version of Costanza *et al.* (1997).

the entire world's population with the present North American lifestyle would need an area of ecologically productive land equivalent to the addition to two extra planet Earths (e.g. Rees and Wackernagel (1994) and some calculations place it as high as ten extra Earths). Such a stance is based on the assumption that supporting people involves the increase in the area of agricultural land and we can only support extra people by the loss of further forest and other areas seen as ecologically valuable. We then additionally learn that the globe is under threat from environmental change, especially climatic change, though the predictions for the exact nature of this future change are often heavily qualified (Houghton *et al.*, 1996a; Houghton, 1997). The "green" (political ecology) orthodoxy which follows has been that the outlook seems bleak unless we act now to limit populations and curb resource use. Quite how we achieve this is not necessarily clear, given different cultural constructs about the world, vested interests and existing infrastructures. The implications are often that society has to change – almost that human nature has to change – for the sake of ecology and the future well-being of mankind.

The history of the green movement has indeed involved the identification of ecological priorities, such as curbing population rise and resource use, and the task is then seen as one of persuading people to adopt such priorities in their lives and work. People are seen as part of the problem and they are exhorted to change in order to adopt ideas and practices which are more ecological. This is not necessarily successful as it can act to set people against ecological ideas because it may involve changes in their lives which may not appear immediately

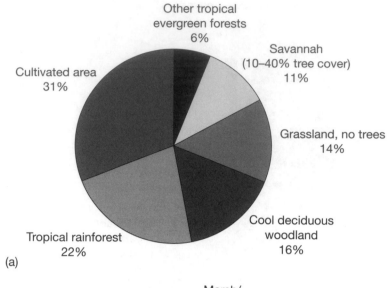

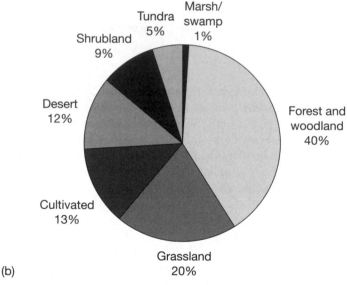

Figure 1.4 World vegetation cover, version of Matthews (1984,1983): (a) 1984, (b) 1983, (c) forest types.

beneficial (I read recently in a newspaper that a majority of Germans now distrust the Green Party there). There is, however, a counter approach which is more positive. Von Weizsacker *et al.* (1997) write of doubling wealth and halving resource use through greater efficiency, which they term a "Factor Four" approach. Tiffen *et al.* (1994) wrote of "more people, less erosion" in Kenya where people are seen as the resource, not the problem – infrastructure and policy changes are needed, not changes to human nature.

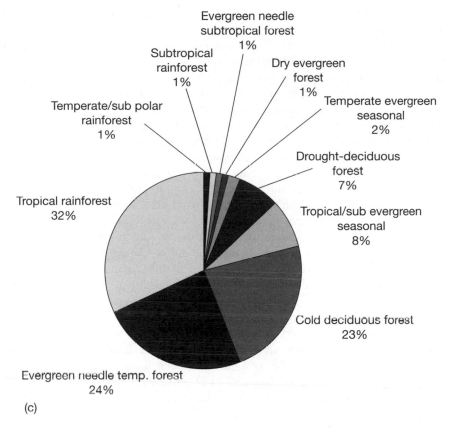

Evergreen needle
subtropical forest
1%

Subtropical
rainforest
1%

Dry evergreen
forest
1%

Temperate evergreen
seasonal
2%

Temperate/sub polar
rainforest
1%

Drought-deciduous
forest
7%

Tropical rainforest
32%

Tropical/sub evergreen
seasonal
8%

Cold deciduous forest
23%

Evergreen needle temp. forest
24%

(c)

Figure 1.4 (continued)

While human nature is itself a wide-ranging concept (Loptson, 1995; Huxley, 1978), and as discussed in *Environmental Psychology* by Bell *et al.* (1996), it is clear that a more modern approach involves one where you are not trying to change human nature nor to coerce people to a particular point of view. In this less crusading approach, people are viewed as part of the solution and ecology takes heed of the nature of society and culture and indeed is seen in the context of serving society. Clearly the notion of "people" covers a wide range of humanity but the approach can be summarized in the phrase: *ecology for people.* This concept admits that people do not necessarily have the environment at the top of their agendas. This more realistic stance does not mean that ecological ideals have to be forsaken but rather that societies do not necessarily have to change radically for the sake of ecology. A stance which invokes societal change as part of the solution to a perception of problems – and a perception which all may not share – can readily lead to the rejection of ecological ideals, or at best lip service or mere tokenism. Rather, it means that ecological ideas can be adapted to a number of different societal structures with a convergence of environmental values, with societies still evolving in their own legitimate way and benefiting from the ecological ideas (Norton, 1955; Minteer and Manning, 2000).

Such a realistic idealism has far more chance of success than a more radical attempt to overthrow or undermine existing orders. Radical postures are still extremely useful, however, as they indicate ways ahead and influence people's thinking. An "ecology for people" is not then a utilitarian dilution of ecological ideals but is a matter of starting from where you are rather than from where you might be, working through people's existing lives and working with real situations (Cernea, 1991), rather than confronting and admonishing people. To achieve this realistic idealism we have to be clear about what ecological ideals there might realistically be. It also includes attempting to identify current society–environment relationships, and how they might vary in the degree of society–environment involvement. The approach first evaluates what we might currently realistically achieve within existing societies and, second, conceives of what we might like idealistically to achieve in the future through social evolution. As an example, it might be realistic not to campaign for the banning of automobiles, which would fly in the face of how many people currently organize their lives, but recognize how much people value mobility, to make cars less polluting and then to minimize their use through steps to improve public transport and facilitate cycle routes.

The psychological barriers to the adoption of new approaches are well documented and so it is not surprising that simple exhortations to conserve and adopt a more ecological lifestyle have not been more widely adopted. For example, writing on the psychology of environmental issues, Sjöberg (1989) observed that if one is proposing to modify people's behaviour by changing attitudes, this will be hard to achieve both because attitudes are notoriously hard to change and because the links between values, general attitudes and behaviour are usually quite weak. In *Environmental Psychology*, Bell et al. (1996) also question whether environmental attitudes predict environmental behaviour. There have been a welter of books which espouse ecological ideals (including, for example, Porritt, 1984; Bunyard and Morgan-Grenville, 1987; Schwarz and Schwarz, 1987; Elkington and Hailes, 1988; Porritt and Winner 1988 and Button, 1989). These green ecological writings might have raised ecological consciousness but many people, while perhaps doing some recycling, still lead substantially unchanged lives in terms of consumption and transport. Indeed, exhortations may be counter-productive and Rowell (1996) writes of the "Green Backlash" and Atkinson (1991) discusses why the political process was so resistant to environmental concerns. Furthermore, Smith (1995) asks the question: "does education induce people to improve the environment?" and finds that merely bringing information to people is not enough – there has to be some pre-disposition or positive inducement. In this context, Bell et al. (1996) feel that attitudes and ethics are actually strongly influenced by direct experience.

Sjöberg (1989) also observed that people maintain a lack of realism, which may indeed even be healthy, so that the good things which may occur seem likely and the bad things as unlikely, strengthening our belief in a safe and stable world with a tendency to disregard conflicting, threatening information. It is also noted that our impressions are slow to change under a belief in perceived stability. Additionally, like stopping smoking, behavioural changes are difficult to make, people are not always rational and do not want to sacrifice time and comfort. It is also interesting to note how people who do, however, actually adopt "ecological" ideas in practice (organic food, cycling to work, for example) derive a somewhat superior feel-good factor of their own. This is possibly in detachment from, and at least partial rejection of, others in society and with a sense of justification from others who share their principles, often via group membership.

Many writers have felt that behaviour change is an important factor in saving the environment, Bell *et al.* (1996) feeling that it will be more important than physical technology in effecting solutions (p. 522). However, it becomes clear that behaviour only changes through experience and the tangible perception of benefit and associations with the feel-good factor. This is traditionally associated with consumption (described by Galbraith (1958) as "the imperative of consumer demand") and with comfort and status, rather than from science and "expert" or pressure group exhortations. This is why the Factor Four approach (von Weisacker *et al.*, 1997) seems more likely to succeed as it proposes to maintain the benefits of consumption while reducing resource use through greater efficiency. Harrison and Burgess, (1994) in their study of people's attitudes to conservation concluded that there was an ecology which uses "a rhetoric that extols the intrinsic worth of organisms" and that this is "based on the assumed universal authority of scientific 'experts' about ecosystem complexity". Confiscatory exhortations from "experts" to change behaviour for the benefit of conservation and the environment could thus be doomed to be unconvincing. This is because the relevant behaviour is more related to perceptions of benefit, vested interest, immediacy, a reluctance to change behaviour and also is based on a simple view of nature – what Harrison and Burgess term a "common sense understanding of how nature is organised" – rather than on science. Additionally, Sjöberg (1989) wrote: "In studies of judgement it has been found that people seldom take more than about five factors into account", yet eco logical science is always emphasizing multiple factors and the complexity of ecosystems and environmental issues (Figure 1.5).

Future benefits of adopting ecological ideals are thus less tangible than people's immediate comfort. If psychologists tell us that people minimize dangers, are slow to change their behaviour and take into account a small number of factors in decision making, then, if there are green exhortations and warnings of disaster on the one hand and scientific postulations of multi-componented complexity, fluid interactions and uncertainty on the other, it is small wonder that few people readily identify, grasp and adopt ecological principles. We are really dealing not only with our own preferences and limitations but also with the limitations in the public understanding of ecological science (Eden, 1996) which guide our conceptualization of the environment.

We live in a world where we seek to make sense of and deal with a multiplicity of possible inputs by finding meaning and order through a number

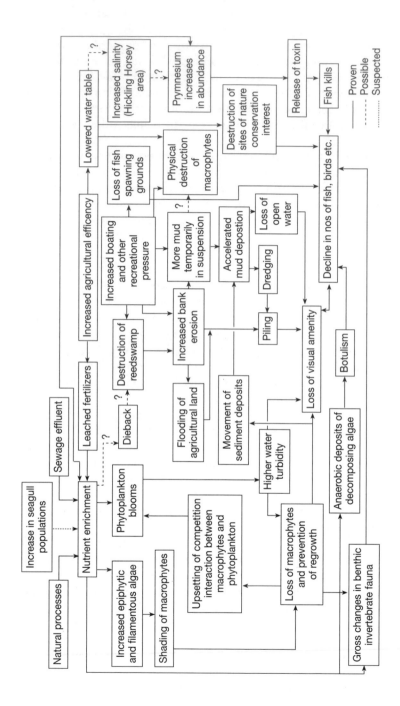

Figure 1.5 Scientists are often emphasizing the complexity of ecosystems. This is a construct of the Norfolk Broads (from George, 1977) showing the relationships between a variety of forcing processes (e.g. sewage effluent and agricultural runoff) and ecosystem response. Many lay people find such constructs baffling. Such expressions of complexity reinforce scientists' expertise and authority but may lead to alienation of people during attempts at public participation – yet scientists find it difficult, in all honesty, to represent the system more simply and remain true to their own constructs of knowledge about ecosystems.

of constructs and concepts, as we discuss further in Chapter 3. Our brains are perhaps better seen as a filter than a receptor (indeed this was the conclusion of Aldous Huxley who, in his book *The Doors of Perception* (e.g. 1977 edition) thought that particular drugs could be taken which reduced this filtering effect, leading to enhanced, but unselective, perception). Such a situation of viewing the world through a number of concepts means that not only do we have the comfort of logical constructs of the world, but also that these constructs act to limit the way in which we see the world. Constructs are ways of organizing our perceptions of the world about us and once we have such constructs they tend to both facilitate and limit our perceptions. One might say that *reality is defined by our concepts of reality* or even that *illusion is the only reality*. If this is so, then it is important to examine our constructs and concepts, as this book attempts to do, because our concepts about ecosystems act to influence the way we treat the world and indeed the way we imagine and manage the terrestrial biosphere. Our concepts can also be modified by the way we perceive how the environment has reacted to our management actions. A quote from the novelist Steinbeck (1951) is apposite here: "A man looks at reality and brings to it his own limitations". In other words, our concepts both facilitate and limit our outlook. We may then be able to justify several concepts about the world and individuals may probably only change their concepts on the basis of evidence which shows that they are not appropriate – or still cling to them despite the evidence.

There have been many concepts of the environment in human history – something fearful, something to be dominated, something to be cherished, something which is degraded by human activity, and so on, all of which have acted to guide the way we approach the terrestrial biosphere. If we adopt the approach that nature can handle anything we throw at it and recover, then, of course we shall be tempted to treat it as robust and resilient, perhaps not without some carelessness. If we see nature as fragile, we might then treat it more carefully. It is possible to gain evidence for any one point of view and it is difficult to be definitive about which social construct might be true as elements of each view might be found (Wiman, 1991; Holling, 1986).

What is certain, however, is that our conceptualization of the environment should include the notion that some ecosystems can be altered by human activity more than others. The logic behind this can be seen as lying in the relationship between cultural energy and environmental energy (Figures 1.6, 1.7). Energetic environments (volcanoes, large rivers, landslides) tend to obliterate the efforts of even the strongest of cultural energies including the artifices and buildings of people. More passive environments (a relatively stable slope formed by slow processes over thousands of years) can easily bear the imprint of human activity such as agriculture, terracing or building. We can thus see that there can be many different combinations and situations in landscapes but they can be defined by the relationship between environmental energy and cultural energy.

I thus believe that it is pointless to now conceptualize the terrestrial biosphere as a physical entity without referring to the past and present effects of human activity. If we can suggest that the terrestrial biosphere is comprised of soils, plants and animals and associated aspects of the atmosphere, hydrosphere and

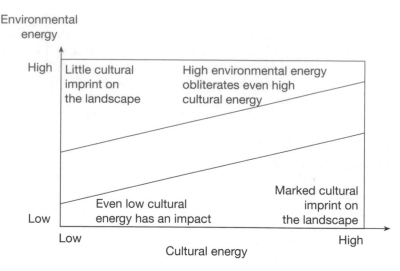

Figure 1.6 Human imprints on the landscape in relation to the interactions between cultural energy and environmental energy. In very passive environments (e.g. tundra, level land surfaces), a low level of human activity can leave clearly defined imprints which last a long time (bottom left of diagram); these imprints increase with increasing cultural energy (e.g. organized teams of earth movement, terracing, wall building), a high level of which can easily obliterate a pre-existing passive landscape (bottom right). As environmental energy increases (e.g. more rapid slope movement, volcanoes, earthquakes, hurricanes – see Figure 1.7) the imprints of lower levels of cultural activity are readily obliterated (top left) and it takes considerable cultural energy to leave a human imprint on the landscape (top right).

lithosphere, then we should immediately also suggest that humans are indeed animals which are constituent part of the terrestrial biosphere. As such, people, just as any other animal, depend upon other components of the biosphere – and can also have considerable effects upon those components. In the case of human beings, these impacts are often seen as greater than those of any other creatures, though when the vast populations of insects and micro-organisms are considered, this may be a debatable point. The key concept is that human beings are not *separate* from the biosphere but an *integral part* of it: they depend upon it and have had an influence on it from within and thus relate to it in a reflexive way.

It would be possible to write a book which considered the ecosystem components including soils, plants and animals other than man and the ecosystem processes involving the flows of energy, nutrients and water – and which then considered the human impact on these ecosystem processes. This would, however, set people apart from nature. While many of us do live in cities physically remote from nature, it is not the case that we are independent of nature or from ecosystem processes. Whatever we do, and wherever we are, we are involved with the environment and, however vicariously, we depend upon ecosystem processes for a flow of commodities, including the air we breathe and the food we eat.

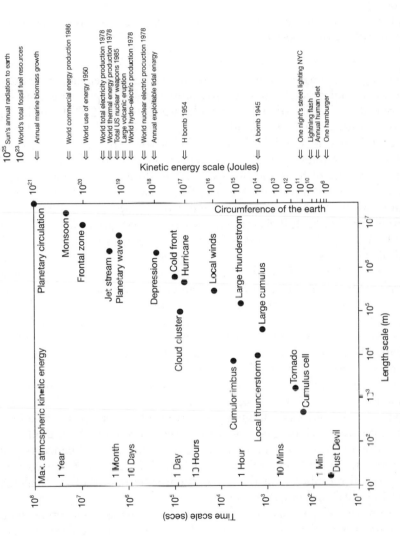

Figure 1.7 A representation of one type of environmental energy. Weather systems in terms of scale and energy (from Barry and Chorley, 1998).

In this context, useful concepts can be derived from the American ecologist, Eugene Odum (Odum, 1989, quoted by Munda, 1997), who divided ecosystems into three categories:

1. Natural/natural solar powered (open oceans, wetlands, rain forests).
2. Domesticated or man-subsidized solar powered (agriculture, aquaculture, woods).
3. Fabricated environments or fuel-powered urban industrial systems (cities, industrial areas, airports).

Munda writes: "it is evident that fabricated environments are not self-supporting or self-maintaining. To be sustained they are dependent on the solar-powered natural or domesticated environments: life supporting ecosystems". This operates self-evidently through trade (see Chisholm (1990) on the separation of production and consumption).

We all depend to a very large degree on the solar energy which is harnessed by plants during photosynthesis, however remote our urban areas may be from the actual source of food and other ecosystem products. *What does vary is the directness of the dependence, not the fact of a large degree of dependence.* Thus, for example, we may harvest and consume our own crops or depend upon others to supply them to us. It is also true that there may be variations in the distribution of ecosystem products to people, due to variations in economic, social and political structures – *the fact of dependency does not vary, but the access to the resources available does.* Thus, despite such variations, people living in any one of Odum's three ecosystem categories have an equal dependence on the terrestrial biosphere – but the directness decreases as you progressively consider people living in the natural through the domesticated to the fabricated states and the access varies geographically and with economic social and political situations. Similarly, the human effects on the terrestrial biosphere vary with place and the cultural and technological level of society.

It seems inescapable, then, that however much we conceive of the science of ecosystem processes as being important, if we are to understand the terrestrial biosphere we should study human perceptions of the environment and social processes as well, including economic and political activity. We should not only look at soil and vegetation processes but also at the cultural concepts and the economic, social and political factors which influence our management of the terrestrial biosphere and thereby influence the nature of our dependency on the biosphere and our effects on it.

A number of authors have attempted to make conceptual pictorial representations of the interrelatedness of the human and environmental systems. The relationships are complex but one of the simpler diagrams is shown in Figure 1.8. Here, three main interactions are envisaged.

- First, external (extra-terrestrial) sources of change can affect both human and environmental systems (e.g. fluctuations in solar activity).
- Second, both human and environmental systems can have their internal sources of change independent of each other.

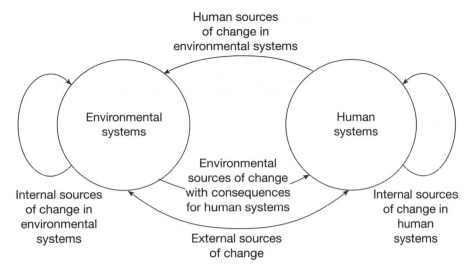

Figure 1.8 Interactions between human and environmental systems (from Clark, 1989).

- Third, there are human sources of change to environmental systems and environmental changes with consequences for human systems.

The third aspect is expanded upon more in Figure 1.9. Here the conceptualization is one where the role of choice in human systems is stressed, with the critical influences of values, options and perceptions which influence our choices about how we might influence and be influenced by environmental change.

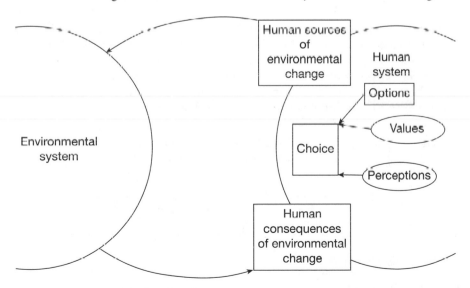

Figure 1.9 The role of choice in reacting to and leading to environmental change. Choice is based on values and perceptions about the options (modified from Clark, 1989).

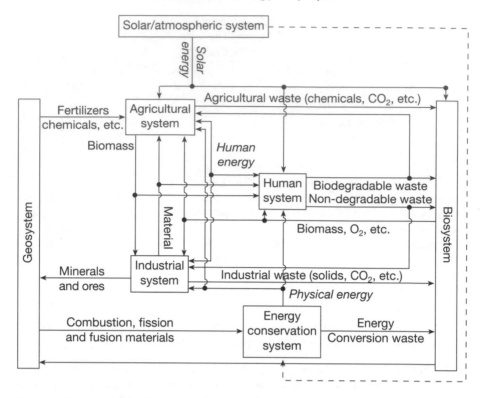

Figure 1.10 A construct of human–environment interactions involving industry and agriculture (from Bennett and Chorley, 1978).

In greater detail, Figure 1.10 conceives of the linkages of material and energy flows between the terrestrial biosphere and human systems involving not only natural but also domesticated and fabricated systems.

For the purposes of this book the situation may be presented as in Figure 1.11. Here the terrestrial biosphere underpins the provision of materials and goods which in turn influences our human value system.

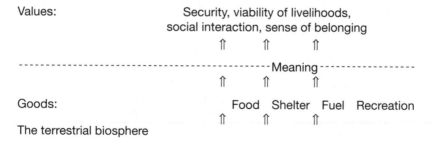

Figure 1.11 The terrestrial biosphere presents items which, through the realm of meaning, are seen as "goods" which relate to our value systems, the provision afforded being not only material but also psychological.

In summary, everyone relates to the environment in some way, but the degree of direct involvement and the directness of the dependency varies. In addition our concepts about the environment are important as they influence the way we wish to interact with it and attempt to manage it.

1.2 People's lives and ecosystems

We often see dramatic pictures on television of drought, famine and war in distant places. But these images appear on our screens and then fade in our memories until some other newsworthy event happens. Behind these headlines, however, is a far wider and unpublicized reality of people continuing to go about their daily lives. It is to this reality that we should turn our attention. What are most people doing most of the time? And how do they relate to the world about them? Some years ago I visited Ethiopia to see an Action Aid project. An extract from my diary helps to illustrate the key points: not a dramatic situation of famine or drought, simply people everywhere going about the daily activities depending upon the land around them (see Plate 1.2). Items that indicate a dependency on, or a relationship with, the terrestrial biosphere are italicized.

Ethiopian diary, Dalocha 12 August 1988

Up early in the grey misty dawn. Hyenas call in the distant hills, a dark slate grey outline with light grey beyond. Gradually the sky lightens over the moist grasses and birds call and chatter. All is very still and quiet; the occasional cock crow. Doves coo, an Ibis wings slowly over, croaking, stark and angular against the sky. Wide flat

Plate 1.2 Dalocha. A rural community in Ethiopia, south of Addis Ababa. Here community needs involve direct contact with the terrestrial biosphere.

Acacias scrape the sky with thorny fingers. The round *mud and wood huts* sog and drip from their *deep brown grass thatch* after the seething evening downpour yesterday when coalescing puddles in the *cultivated fields* quickly joined into torrents, *taking the brown red soil with it* as it flowed. People cough and stir and the first smoke drifts and wreathes into the sky as the cattle are let out – *munching and tearing at the grass* with loud rasps and snorts; *women clear the dung from the night before and pack it to dry.* A suckling baby murmurs from the shadows. Black wet blobs of grass heads dry in the sun and diffuse into a dotted haze as they dry. The tall grasses poise into the sky like a shower of spears the cattle guards use. Patches of lighter green appear at the tops of the eucalyptus trees as the sickle-shaped black sky-cut leaves become green as the sky lightens. *Fires are lit and the wood crackles* as the pungent aromatic eucalyptus resin smell drifts while people fan the *charcoal.* Someone, somewhere *chops wood* and sounds echo into the greyness. A figure shrouded in a blanket regards the chill moist air and walks slowly from a hut *to gather water from the spring.* Women set off for the *distant hills to gather firewood* – a five-hour round trek.

The key elements, italicized, here are security, food and warmth – and show the relationship with the terrestrial biosphere. The huts provide security – they are made of wood, mud and grass for thatch: settlement in the area had inevitably involved some deforestation for the poles which support the roofs of the huts. The crops and cattle provide food – these depend on the soil, the growth of grass and the planted crops. Firewood provides warmth and heat for cooking – the wood comes from the forest and the planted eucalyptus trees. There is human need to be able to sustain such a provision.

The facts of being able to sustain such a provision are simple:

1. Harvest vegetation at a rate which matches natural re-growth, grow crops on the soil in a way which conserves the soil and extract water at a rate which matches natural replenishment by rain water and you have infinitely sustainable resources – and thus also a society which can be sustained.
2. Extract or lose soil, forest and water resources faster than they are replenished and the resources cannot be sustained; the well-being of a society which depends on those resources is undermined.

The elements of some of the problems, such as soil erosion, which I saw in Ethiopia lay to some extent in the politically motivated settlement of people in one area. In some areas, people had previously been nomadic, moving with the rains to fresh pastures where they occurred and thus being "in tune" with what nature could provide in different areas at different times. Settlement made them dependent on one area and on the resources it could provide all year round – a building of settlements naturally increased the use of local timber for housing, leading to a reduction of forest resources round the settlements. The elements of success in the Action Aid project I was visiting included local plantations of trees for fuel (to save a five-hour trek), building of small reservoirs to conserve water resources and soil conservation measures. But these alone were technical solutions: the real success lay in the participation of the people – and the fact that they are individuals, with their own legitimacy, not just some idealized and abstract notion of "the people" (see Inset 1.1). There existed considerable local

---

Inset 1.1   Participation

'What gives real meaning to popular participation is the collective effort by the people concerned to pool their efforts and whatever other resources they decide to pool together, to attain objectives *they set for themselves*. In this regard participation is viewed as an *active process* in which the participants take initiatives and action that is stimulated by their own thinking and deliberation and over which they can exert effective control. The idea of passive participation which only involves the people in actions that have been thought out or designed by others and are controlled by others is unacceptable.'

Source: from ACC Task Force, Working Group on Programme Harmonisation, Rome 1978 quoted by Oakley, P. and Marsden, D. (1984) *Approaches to participation in rural development*. International Labour Organisation (International Labour Office, CH-1211, Geneva 22, Switzerland).

---

knowledge about soil, plants, crops and livestock which had only to be put into practice. But why then, had any problems such as soil erosion and the loss of forest arisen? In many ways, these were related to political practices, like the resettlement schemes mentioned above, rather than to any ignorance of the people involved.

This illustrates the fact that knowledge about ecosystems, while fundamental, is alone not enough. *The socio-economic and political infrastructure also needs to be appropriate if we are to use our resources in a sustained way*. In fact, in many instances it is these factors of human organization which give rise to environmental problems and the inappropriate use of resources rather than necessarily from any ignorance of ecosystem processes. This does not immediately give us any answers as to what is "appropriate" but this is to suggest that a lack of ecosystem knowledge is not so much an impediment as having inappropriate social institutions might be (see O'Riordan and Jordan (1999) for a discussion on the role of institutions and see Honadle (1993) on institutional constraints).

But what of people in cities, far removed from the land – Odum's fabricated environments? The same principles apply, the difference being the transport of the food and often woodfuel to the city; there is the same dependency on ecosystem resources even if the relationship is less immediate. We cannot see the *immediacy* of the dependence in the "Ethiopian diary" extract above, but think about your daily life and a dependence on the biosphere will still emerge there too. This makes an equally good classroom or individual project. For me I sleep in a bed made of *wood* in *cotton* sheets in a room in a house where the roof is supported by *wood*. Breakfast sitting on *wooden* chairs involves *agricultural products*: wheat, margarine, oranges in the marmalade, coffee; *energy* use in cooking, lighting and heating and the mechanized means of transport produce *carbon dioxide*, work involves reading and writing words on *paper* and shopping and eating involves *meat, vegetables* and *fruit* from many parts of the globe.

Evenings might involve watching television programmes about the world about us. Thus, while there are, of course, alternative products in many cases which need not have so much of a dependency, or an impact on, the biosphere, the themes here are ones of consumption and of goods brought to the individual by various means of transport and of energy consumption, indeed a dependence upon and impacts on the terrestrial biosphere together with, through television, a vicarious involvement with nature. The only direct involvement may then be through gardening and recreational walks in the countryside, much valued for their therapeutic effects (see Chapter 5).

The central points are that in order to ensure that there is an environment which we can appreciate, and to promote human well-being, we need to understand how people relate to the terrestrial biosphere and how ecosystems function in order that we can use them in a sustainable way. We also need to understand how societies organize themselves and interact with their environment before we are able to put this understanding to good use.

1.3 Involvements with ecosystems

All societies are dependent on, and have an influence upon (see Inset 1.2), ecosystems, but the degree of direct involvement and the dependencies vary. A number of situations can be identified.

Minimal human influence

1. Wilderness areas. These are areas which are inhospitable for human life and which thus have low population density with low human influence, but may be managed to a certain extent and may probably have been influenced by human activity in the past. The wilderness element may thus be more imaginary than real. Included are ice caps and other polar regions, mountain ranges, deserts, steppes, remote forests (both cold northern and tropical) and other areas where *human activity does not dominate over environmental conditions*. At the simplest level of direct human dependence are the hunter-gatherers, who (while there are not many left in the world) are dependent on what the ecosystem provides. These people have a very direct involvement with the ecosystem in which they live. They are often thought of as having very little impact on the ecosystems in which they live: human activity can be seen as adapted to the ecosystem, though the notion of a "harmony" between "primitive" societies and nature can be contested and there have been widespread influences, such as that involving the use of fire. The involvement with human activity may involve elements of subsistence or low intensity agriculture or small settlements. Another dependence is for recreational activity and spiritual reassurance for people from other situations, notably from areas of agricultural and urban development. Some commodity extraction (e.g. timber) may occur, but with little impact on ecosystem structure and function. Such areas may be declared as wilderness areas and afforded some degree of protection and indeed be subject to some

---

Inset 1.2 Disturbance of ecosystems

In terms of disturbance, Hannah *et al.* (1994) have classified ecosystems into undisturbed, partially disturbed and human dominated:

- Undisturbed: primary vegetation, very low human population density (under 10 persons per km^2 or in arid/semi-arid and tundra communities under 1 person per km^2).
- Partially disturbed: shifting or extensive agriculture, secondary but naturally regenerating vegetation, livestock density over carrying capacity or other evidence of human disturbance (e.g. logging concessions).
- Human dominated: permanent agriculture or urban settlement, primary vegetation removed, current vegetation differs from potential vegetation, desertification or other permanent degradation.

They estimate from studies of land cover that each of the three categories occupies around a third of the earth's habitable surface:

Total area of world habitat: 162,052,691 km^2.

Of which:
Rock, ice and barren land: 27,148,220 (16.75%)
Remainder 134,904,471 (83.25%)

Of this remainder of habitable land:
Undisturbed: 36,424,207 (27%)
Partially disturbed: 49,509,941 (36.7%)
Human dominated: 48,970,323 (36.3% or 23.9% of total world area)

Wilderness ecosystems (1) (see main text) will correspond to Hannah's Undisturbed category; managed and subsistence agricultural (2 and 3a) will correspond to the Partially disturbed category and Low intensity agriculture, intensive agriculture, rural settlements and urban ecosystems (3b,c, 4 and 5) to Human dominated.

---

degree of management or they may be undeclared; in either case management intervention is minimal. The strongest influences may be sporadic (though having great effect if repeated, e.g. burning of grassland or forest) or indirect (as in the case of air pollution, for example, where industrial pollutants rise in the atmosphere in northern temperate regions to be deposited in the arctic in cold, downward flowing polar air).

Modified ecosystems

Here, the ecosystem is *adapted to human need* and there is some element of species change, either in terms of percentage composition or of type and in age structure, form and also ecosystem function, including water, energy and nutrient cycling. There are also physical manipulations of the environment.

2. Managed ecosystems. These are ecosystems which may be (a) nature reserves which are putatively natural but are kept as they are by conservation management intervention, or (b) managed forests which may be planted or otherwise managed and used for forest products.

3. Agricultural and other extractive systems. Here the ecosystems are modified for the purposes of growing food, timber and other ecosystem products (including agroforestry systems).

(a) Subsistence. Here people grow their own crops and while they may also hunt and gather they are not entirely dependent on what the ecosystem provides as they have their own managed and modified ecosystem. The crops and livestock are used by the community and imports and exports from the community are minimal. The impact on the ecosystem involves extraction (timber, fodder), clearance for cultivation and grazing. There is *very little trading of ecosystem products.*

(b) Low intensity agriculture. Here people are involved in an artificial ecosystem, dependent on the soil resource and planted crops with *an element of off-site trading.* The impact here is a new agricultural ecosystem but elements of non-managed ecosystem and modified ecosystems may also exist. In contrast to (3a), inputs of capital and materials and outputs of products exist but they are low relative to (3c).

(c) Intensive agriculture. As (3b) but inputs of capital and materials (fertilizers, pesticides, seeds) and outputs of products are high relative to (3b). The system is *geared to off-site trading.* This category grades into the next: Rural settlements.

Ecosystems replaced

4. Rural settlements. Here we are considering westernized situations where *direct dependency on the ecosystem is lost* and the occupants depend on others (3b, 3c or forests) for ecosystem products (rural settlements in other situations can be considered under 1, 2 and 3a and 3b). The involvement with the local area is locational rather than necessarily functional (e.g. dormitory villages in rural areas). The impact on the ecosystem (soil and vegetation) is that it is replaced by something which is controlled or permitted, including gardens. The ecosystem experience is often limited to recreational activities and experiences of landscape preferences.

5. Urban. There is neither a direct dependency nor a locational involvement; occupants depend on (3b, 3c or forests) for ecosystem products. Apart from the atmospheric elements and weather, the *ecosystem is controlled*, including parks and gardens, permitted or suppressed. The relationship with the environment is largely through the media, especially television and involves a series of abstract notions as presented rather than any direct, daily involvement. Direct involvement is largely confined to vacations, through tourism and through recreational activity.

Thus one moves from low impact environmental dominance to high impact environmental control. The degree of dependency is equal in all cases but the directness decreases as the degree of trade increases.

1.4 Ecology, social justice and sustainability

In an issue of *New Internationalist* (No. 307, November, 1998) highlighting Eco-socialism under the heading "Red and Green" Richard Douthwaite proposes that "equality and sustainability can't do without each other". He writes: "many Greens might feel that aiming for equality and sustainability is simply taking on too much. Achieving either target, they might say, is well-nigh impossible by itself, but setting out to achieve the two together is simply ridiculous. Such a view is, however, seriously mistaken. The two goals are inextricably linked because the least environmentally damaging way of providing any given level of human well-being is for everyone to get more or less the same. Equality, in short, is ultra-efficient." He concludes: "Equality isn't a sideshow, a distraction in the quest for sustainability: it's simply the best way there".

Whatever one's political persuasion, it is clear that it can be argued that the well-being of a society and the state of its environment are inextricably linked. Indeed, it seems quite tenable to propose that we should be striving to improve both simultaneously through an equitable distribution in society of the benefits of an ecologically sustainable environment in what could be termed the "win-win" situation. Indeed, Di Chiro (1995) writes of the *convergence of environmental and social justice*. However, we first need to establish what we might mean by terms such as sustainability and examine our assumptions about social–environmental interactions from first principles.

All living organisms use energy. They manufacture their components from chemical elements in their surroundings in the presence of water utilizing the energy from chemical reactions, the sun or from ingesting other organisms. It follows that life is dependent on water, energy and chemical elements in the environment and that life forms depend upon, and have effects on, their surroundings. Ultimately the components extracted from their surroundings return to the environment through death and decomposition. Animals cannot use the sun's energy to create their components but are dependent on ingesting plants which can do so: animals are dependent on plants.

Organisms survive for longer if they can find protection from adverse environmental conditions and other predatory organisms, either though physical protection, physiological adaptations or behaviour, including evasion tactics and also storage of food. In the face of such pressures, a great diversity of organisms have arisen, commonly seen through Darwinian theory as the result of genetic mutation providing advantage for survival, or at least no disadvantage.

Human beings are no different in these respects from any other organism, with the need for energy, water and chemical elements; a dependence upon and influence on the environment and in seeking protection and showing behaviour which includes food storage.

A critical difference is that human beings display a high level of consciousness – they are aware of themselves and of their surroundings; they can also remember the past and envisage the future. Such characteristics can be found to some extent in other animals, such as primates, and so consciousness is probably a matter of degree rather than being a uniquely human trait but being displayed by human beings most markedly. People are certainly highly aware of their vulnerability and this insecurity is marked in many societies by an increase in the trait of acquisition and possession to give them security (though this can vary, as Laurens van der Post (1974) observed "the Europeans specialise in having but the Bamuthis in being"). They also are able to articulate and feel social position, often in relation to these possessions. They have both the sense of belonging to social groups and places, which may involve the implicit rejection of other groups or places, and the notion of success relative to others, often expressing itself in terms of competition. These social attributes are also displayed, though not necessarily consciously, to some degree by other organisms but they are most marked in, and uniquely articulated by, human beings.

In addition human beings are able to make conceptualizations of things other than themselves, to act self-consciously in altruistic ways and also display the characteristic of a belief in something higher than themselves. In part such beliefs fill the need for certainty which arises from the consciousness and articulation of uncertainty. They also possess the concept of right and wrong.

These values change the nature of the relationship to the environment. There is not a passive relationship but a reflexive active evaluation of benefit. There is also a marked tendency to feel secure through acquisition together with recognitions of social standing. There is also a range of spiritual relationships with the environment.

Environmental issues might then be seen as arising through conflicts between (a) acquisition and possession, with its attendant social standing, and (b) the realization that this acquisition may impair current or future acquisition of goods or territory. Here, there is a threat to continued well being perceived by at least some people or groups. Indeed it is often the actions of some groups which are perceived as problematical by other groups which give rise to issues based on conflicts of interests (as discussed by Trudgill (1990)). Also involved is the spiritual notion that human beings have altered states which existed prior to their effects on the environment with the attendant judgements the desirability of these alterations, together with the ethical notions associated with The Garden of Eden, guilt and Utopia. The essence is that we act as a tribal species where groups, acting out of their own self-interests, may differ in their aims but also that we may act together when there is a perceived general threat. In this sense we are no different from other species which show co-operation between individuals for the common good. However, the difference is that the human species is differentiated into nations, classes, consumers, producers and by wealth, education and locality and such a differentiation makes co-operative effort difficult because it involves overcoming loyalty to such groups.

Alterations to the environment and to other organisms are thus no more than the effects of behaviour which are displayed by other organisms and should be understood as such – human needs are the same as the needs of other organisms – food, shelter, security. *Unless we understand the basic needs for acquisition, social standing and social differentiation we will not progress with environmental issues* – many who espouse green issues do so in a way that manifests itself as confiscatory and punitive, appealing only to abstract ideals and spiritual values while ignoring social needs and differentiation. This palpably does not work. Where fundamentalist "green" ecologists failed is that they did not allow for human characteristics, seeing them as greed, profit or exploitation in an otherwise negative sense. Profit and gain are not somehow lower values but are basic values which are human characteristics. This does not condone every human action which destroys the environment as pardonable in the context of acquisition but it at least helps us to understand what is going on: we have to understand that we should provide food, shelter, security and social standing while simultaneously conserving the environment and other species for our own benefit and for the sake of our spiritual values which sees the environment and other species in their own right. We also have to do this in a way that is sensitive to social differentiation in terms of values and of willingness and capability to act. The actual problem may then be that some people have more than others – "more than they need" – in other people's views. The quality of life issue is important in this consideration and O'Riordan (1992) suggests common components in quality of life definitions (see Inset 1.3). Calls for equity have to consider whether it is about raising standards for all to the highest existing or more of a levelling effect to the lowest. Achieving some intermediate point or accepting differences in acquisition has always been a challenge for societies.

Human needs have been described by Kekes (1994): "Many of these needs are physiological: for food, shelter, rest and so forth; other needs are psychological: for companionship, hope, the absence of horror and terror in one's life, and the like; yet other needs are social: for some order and predictability in one's society, for security, for some respect and so on." These needs are referred to as "primary values" in contrast to "secondary" values which derive from the satisfaction of needs that vary with traditions and conceptions of a good life. The fundamental challenge is seen as lying in defining a good life relative to the resources available: As we noted earlier (section 1.1) calculations suggest that to provide the entire world's population with the present North American lifestyle would need increases in the area of productive land. The counter conclusion might be that we have to reduce the consumption of Earth's resources by western nations in order to meet a view of social justice that requires the equal distribution of benefits to all. However, Smith (1997) argues that an egalitarian formulation of distributive justice requires "*reference to the way of life in which it is embedded*", with an emphasis on diversity and difference. This is a common theme and, for example, England (1994) in a study of women in cities also stresses that urban social justice needs to be sensitive to difference and diversity. Gallopin and Öberg (1992) agree that quality of life is by definition a subjective concept dependent on cultural perspectives and values. This really means that

Inset 1.3 Quality of Life definitions

O'Riordan (1992) suggests common components in Qualify of Life definitions:

1. BASIC NEEDS
- human survival
- health
- warmth-energy
- shelter
- water
- nutrition
- personal security

2. BASIC SOCIAL SERVICES
- education
- medicine

3. ECONOMIC ASPECTS
- means of livelihood
- employment
- income
- material wealth

4. COMMUNITY CHARACTERISTICS
- shared value system and norms of behaviour
- social support network
- participation in group decision making and action

5. ENVIRONMENTAL QUALITY
- sustainable natural resource use
- pollution and degradation
- means for recreation
- survival of other species
- vicarious satisfaction

6. TIME AND SPACE DIMENSION
Time
- positive view of the future
- knowledge of individual and collective past history
Space
- identify with home and territory
- freedom of movement

social justice involves distributive equality within specific site and societal contexts. Certainly there are many calls for policies to be calibrated in a site-specific way (Trudgill and Richards, 1977). Both Poore (1989), when discussing the sustainable use of forests, and Carpenter see sustainability as referring to sites: Carpenter (1996): "My working definition: sustainability is when the productive potential of a managed ecosystem *site* will continue for a long time under a particular management practice." There is also the call for national policies to promote local empowerment (Agarwal and Narain, 1990) and Wilson and Bryant (1997) present a useful construct of the levels at which this might work (Tables 1.4(a) and 1.4(b)).

Many writers do not see economic considerations of the productivity of ecosystem sites as opposable to environmental concerns: the "win-win" situation. Munda (1997) feels that "sustainable development carries the ideal of harmonisation or simultaneous realisation of economic growth and environmental concerns" and quotes Barbier (1987) in that sustainable development implies that we should endeavour "to maximise simultaneously the biological goals (genetic diversity, resilience, biological productivity), economic system goals (satisfaction of basic needs, enhancement of equity, increasing useful goods and services) and social system goals (cultural diversity, institutional sustainability, social justice, participation) to which I would add social and personal goals and values.

Table 1.4(a) Key types of environmental managers (from Wilson and Bryant, 1997); EM = Environmental Management

Environmental managers	Examples	Level of environmental interaction
State	Department of the Environment Department of Forestry Ministry of Agriculture	Actively and self-consciously manage the environment at the local, national and global levels
Environmental NGOs	Greenpeace Friends of the Earth Haribon (Philippines)	Active and self conscious role in influencing decisions about EM at the local, national and global levels
TNCs	Rio Tinto Zinc Matsui Siemens	Actively and self-consciously manage the environment at the local, national and global levels
International financial institutions	World Bank International Monetary Fund Asian Development Bank	Active and self-conscious role in influencing decisions about EM at the local, national and global levels
Farmers, fishers, nomadic pastoralists, shifting cultivators	Farmers in the UK Spanish fishers Buroro nomads (Africa)	Actively and self-consciously manage the environment at the local and regional levels
Hunter–gatherers	Penan (Malaysia) Yanomami (Brazil)	Actively and self-consciously manage the environment at the local level

Table 1.4(b) Market integration and multi-layered environmental management (from Wilson and Bryant, 1997)

	Role in the market	Potential power in the market	Degree of market integration/resiliency	Response to scarcity and externalities	Associated predominant EM practice
TNCs	• Producers	• Price setters • Price shapers • Price receivers	• Strong integration • Variable resiliency (depending on size, location, etc.)	• Contribute to scarcity and externalities • Seek to avoid regulation of activities contributing to scarcity and externalities	• Maximum exploitation • Little regard for conservation
State	• Producer • Direct regulator	• Price setter • Price shaper • Price receiver	• Strong integration • Usually resilient	• Contributes to scarcity and externalities • Seeks to regulate (directly/indirectly) scarcity and externalities	• Maximum exploitation • Some (and growing) regard for conservation
International financial institutions	• Indirect and direct regulators	• Indirect price shapers	• Strong integration • Usually resilient	• Indirectly contribute to scarcity and externalities • Indirectly seek to regulate scarcity and externalities	• Promotion of maximum exploitation • Growing attention to conservation issues
Hunter–gatherers	• Producers	• Price receivers	• Weak integration • High resiliency	• Adversely affected by scarcity and externalities	• Increasing pressure to alter often conservation-oriented practices
Farmers	• Producers	• Price receivers • Price shapers (e.g. large farmer lobbies)	• Moderate to strong integration • Low to moderate resiliency	• Contribute to scarcity and externalities • Adversely affected by scarcity and externalities	• Often forced to maximize exploitation
Environmental NGOs	• Indirect regulators	• Indirect price shapers	• Weak integration • High resiliency	• Indirectly seek to regulate scarcity and externalities	• Promotion of maximum conservation consistent with social justice

In the context of forest management, Mendelsohn (1996) writes about an economic–ecological model for ecosystem management and cites the definition of the purpose of forest management by Gifford Pinchot: the purpose of forest management is to provide "*the greatest good, for the greatest number, for the longest time*". In this view, all forest services are to be "counted" in, not just services sold in markets; the resources should be operated for all the public, at least the public resources, that is the values of all people who care about the forest should be included. The consequences of the decisions should also be included in the decision-making process. Mendelsohn (p. 214) summarizes the view in mathematical terms:

$$V \propto Q + X$$

where:

V = value
Q = market services (saleable produce)
X = non-market services

The calculus should then maximize the present value of the services, both in the short term and the long term.

Market services, Q, includes sawn timber, pulp, grazing and mining while non-market services, V, include hunting, fishing, water flows and recreation with less tangible but nonetheless important carbon storage, endangered species and existence value. This is an important stance as it sees ecosystem goods in terms greater than simple forest products and recognizes the importance to people of the less-tangible products.

Many writers have rehearsed the tragedy of the commons, where exploitation of a common asset is inevitable. This may be the case in a capitalist context, but use of a common asset also contains the possibility of greater equity in the use of ecosystem products provided the social organization is appropriate. Some writers on ecology and social justice have concluded that a Marxist approach is a viable option (Pepper, 1993; Harvey, 1996) but, in the realization of the diversity of the societies we have on earth it is perhaps more fundamental to say that "green development programmes must start from the needs, understanding and aspirations of individual people, and must work to build and enhance their capacity to help themselves" (Adams, 1990, p. 201). This means that they must have enough appropriate information and be empowered to put it into use, a processes which involves not only education and bringing information to people but also facilitating the appropriate practices. This is the conclusion that Trudgill (1990) reached, one which Johnston (1996) endorsed and one which, for example, the Ladakh Ecological Development Group have tried to put into practice, as described by Bunyard and Morgan-Greville, (1987, 267–272). But, having sketched out some of the basic ideas as to the way forward, do we have the "appropriate" information? And how can the appropriate information be meshed with a realization of the legitimacy of social agendas and political power? Trudgill (1999) sees the way forward as progressing from the older "green" agenda to a more realistic stance:

Ecology v. people

---

1. 'Greens' – identifying ecological priorities and persuading people to adopt them.
2. People are seen as part of the problem – asked to change to adopt ecological ideas.
3. Successful? – acts to set people against ecological ideas because of the changes.

Ecology for people

---

1. People and their varied agendas viewed as legitimate and part of any solution. Allows for human nature and admits that people do not necessarily have the environment at the top of their agendas.
2. More realistic? – not forsaking ecological ideals but does not necessarily mean societal change for the sake of ecology.
3. Ecological ideas adapted to and maximized within different societies still evolving in their own legitimate way.

Further reading

**   Key reading
*   Important reading
Recommended reading

Chertow, M.R. and Esty, D.C. (eds) (1997) *Thinking Ecologically: The next generation of environmental policy.* Yale University Press.

Houghton, J. (1997) *Global Warming: The complete briefing.* (2nd edn). Cambridge University Press.

Kemp, D. (1994) *Global Environmental Issues* (2nd edn). Routledge.

*   Munda, G. (1997) Environmental economics, ecological economics and the concept of sustainable development. *Environmental Values*, 6, 213–233.

Proctor, J.D. (1995) Whose Nature? The contested moral terrain of ancient forests. In Cronon, W. (ed.) 1995a. *Uncommon Ground: Toward reinventing nature.* W. Norton.

*   Rowell, A. (1996) *Green Backlash: Global subversion of the environmental movement.* Routledge.

**   Sack, R.D. (1990) The realm of meaning: the inadequacy of human–nature theory and the view of mass consumption. Chapter 40 in Turner, II, B.L., Clark, W.C., Kates, R.W., Richards, J.F., Mathews, J.T. and Meyer, W.B. (1993) *The Earth as Transformed by Human Action: Global and regional changes in the biosphere over the past 300 years.* Cambridge University Press, 659–671.

*   Smith, D. (1997) Back to the good life: towards an enlarged concept of social justice. *Environment and Planning D: Society and Space*, 15, 19–35.

Wilson, G.A. and Bryant, R.L. (1997) *Environmental Management: New directions for the twenty-first century.* UCL Press.

Chapter 2

Are there any guiding principles from ecological science?

Summary of key points:

1. Both science and policy have their own agendas and there is no necessity in the relationships between them.
2. The spectrum runs from perceptions of science-led to actually policy-led situations.
3. Science can have three meanings: internal, self-defining (scientific) ones, generalizations (for the subject); and external. It is this external, culturally-defined significance which is crucial.
4. There are perceptions of linkage, interaction, flow, cycling, adjustment and reaction in ecosystems but, beyond that, perception of translations of "ecological principles" into the management arena are seen as involving value judgements and preferences rather than any necessity.
5. Science has a great deal to contribute in the simpler, mechanical, verifiable systems, but in the "loose", less tangible and unverifiable constructions like ecosystem and landscape, science is a foundation but it has little to offer in the sense that it can't decide what views people will hold and how they will act.

2.1 Science and environmental decision making

Science never proves anything but goes by something as long as it works well.
Steinbeck (1951) *The Log from the Sea of Cortez.*

Natural science does not consist of ratifying what others have said, but in seeking the causes of phenomena.
Albertus Magnus (?)1193–1280

We are arguing as to what would happen in the event of a hypothesis which has not been verified by our experience. The principle involved is ... in the absence of evidence to the contrary, we shall assume the unobserved portions of the universe to obey the same laws as the observed portions. But unless we have an independent definition of truth concerning what is observed, this principle will be a mere definition, and the "unobserved portions" will only be a technical device, so long as they remain unobserved. The principle only says something substantial if it means "what I shall observe will be found to resemble what I have observed", or, alternatively, if I can define "truth" independently of observation.
Bertrand Russell (1940) *An Inquiry into Meaning and Truth*

In *The Western Intellectual Tradition,* Bronowski and Mazlin (1963) wrote that "An early basis for science had been found by Thomas Hobbes (1588–1679) in the principle that *every effect has a material cause – and the same cause always has the same effect*" (emphasis added). Later Hume (1711–1776) was to argue that "*necessity is something that exists in the mind, not in objects*", which he thought attacked the notion of causality by demonstrating that nature gave no evidence for it. He thus felt that there was no empirical evidence for causality. This was something which Kant (1724–1804) took to actually support his view that nature must conform to causality, which he held because he thought that causality is the only way in which the mind can grasp nature's workings. Kant concluded that empiricism was not enough to display the underlying causality which *must* exist and that nature can then only be understood if we see that underneath the (inadequate) empirical experience lies a framework of *a priori* knowledge.

Bronowski and Mazlin's writings have resonance today. First it shows that from an early stage the same proposals can be interpreted in different ways by people with different assumptions, Hume meaning to deflate causality and Kant taking his words as supportive of causality, having decided that it must exist. Second, that while in 1995 Pahl-Wostl, when considering ecosystem science, could write, as we saw in Chapter 1: "Scientific knowledge does not simply emerge in the brains of ingenious researchers enlightened to reveal the truth in nature. Knowledge must be seen from a relational perspective depending on the values, beliefs, perceptions of nature and society", many people in ecosystem management contexts are still asking for, and expecting, "scientific evidence" – or simply knowledge of cause and effect relationships from an "objective" stance where the "truth" is revealed. In 1987 Bird wrote in *Social Construction of Nature* that all you can claim is "the relative truth about nature, one whose meaning is governed by a particular scientific paradigm" and Williams (1994) quoting this in his discussion of environmental history concludes that "not only ecology but all science is a social construct and that there is no objective reality on which to draw". Finally, Cronon (1994), in a critique of Demeritt's (1994) discussion of historical ecology makes a statement which perhaps reflects current thinking: that we can never "know nature … with an absolute, objective certainty … I believe that our knowledge will always be partial and heuristic, best tested by pragmatic consequences". He does, however, go on to discuss the suggestion that Demeritt appears to make that "*descriptions of nature conform to natural reality only insofar as they produce that natural reality*", saying that while such social constructionsim (see Demeritt, 1996) tells us something important about the real world, it is not everything and: "the … redoubt of realism is not quite so ruined" (and see Bhaskar, 1975).

Dunbar (1995), in his pro-science book, rehearses the feeling that theories are what we use to make sense of what we see and serve to direct our attention to particular phenomenon, concluding, after much discussion, that "*science, then, is a process of intense criticism*". It would seem that while we might conclude that all knowledge is positioned, the role of evidence is still a real issue as it was for some of the earlier philosophers of science, albeit that we now realize that it might be supporting a preconception rather than indicative of the "truth". We may then find that the notion of "warranted assertibility" from Russell (1940) is still yet useful in terms of a (positioned) statement which is a justifiable conclusion – and a conclusion relative to the justification (and where the justification can be contested). We might then say that both chaos and order are illusory and that the only certainties are derived from mutual definitions: the sun rises every day – the day is defined by the sun rising.

It is notable, however, that in the media, while there is a growing mistrust of "expert opinion", science can still be represented as an authority for management action – and this even though there are many constructs about science and the relationship between science and management. While Newton (1997) has written of *The Truth of Science*, many writings have indeed challenged the assumed authority of science. Lindsay (1995), for example, feels that science is more about dealing with the unknown than investing in what it is thought to be known. The author writes that people continue to believe in the science of certainty and the certainty of science but he feels that it is entirely unreasonable for it to be expected that science should come up with precise answers: "The philosophy of scientific certainty and the fear of ignorance have brought about a general belief, not least in the minds of developers and politicians, that the scientist merely has to wave a magic wand over the problem for all the answers to pop out with absolute precision ... this is rarely the case with the environment." Indeed, Lindsay was writing in a book by Wakeford and Walters (1995) entitled *Science for the Earth* and subtitled *Can science make the world a better place?* A response to this question could be "But why should it?" Science has its own agenda which need not refer to applications and indeed in that book Lindsay (1995) argues that science should be more concerned with the unknown rather than applying what is known. There may also have been temporal shifts in the role of science in policy making (see O'Riordan (1995) *Environmental Science for Environmental Management* for an earlier discussion). In North America, and in the context of river basin management, Graf (1994) suggests that these categories of the interrelationship may have had the sequence, in terms of shifting agendas and development of methodology, as follows:

1. 1900–1930. Policy without science.
2. 1930–1950. Politics using science for environmental and social control.
3. 1950–1970. Period of experimental science.
4. 1970–present day. Period of environmental impact assessment.

This example, for western American rivers, suggests that the importance of science has grown over time, especially since 1950. However, Eden (1998) now comments that "environmental knowledge is a product not merely of scientific practices but also of research cultures and of negotiations between science and policy, between science and its publics and between claims to authority. Social constructivist studies of science ... have pulled apart notions of rationality, knowledge and certainty ... people ... depend on 'second-hand nonexperience' ". There are thus a range of views and many things to unravel in terms of cause and effect, certainty, public opinion of science, the cultural and social positioning of science and knowledge and the scientific basis of decision making.

It is then perhaps a false dichotomy to polarize a complex relationship into two extremes, but nevertheless such an exercise is useful as a starting point to clarify the issues involved. The two extremes of the relationship between science and policy making are that:

1. science presents its results and they then become available to policy makers; or
2. policy makers decide policy on political grounds and they may (or may not) then refer to science subsequently. In this context, Boehmer-Christiansen (1994) argues that "research institutions tend to produce ambiguous advice while politics will use scientific uncertainty to advance other agendas".

Such a polarization between science-led decision making and policy-led decision making is but a starting point for the discussion in that it assumes that both science and policy making are homogeneous activities; the reality is that there is a complex dialogue between science and policy making, and both can vary in their agendas:

1. Policy making might ignore, manipulate or be based on science in identifiable categories which are not necessarily mutually exclusive:
   (a) it may ignore science and operate according to *political priorities*; or
   (b) it may utilize scientific "evidence" to *reinforce a political decision*; or
   (c) it may hide behind evident complexity in order to *justify inaction*; or
   (d) be exasperated by complexity when *seeking a clear mandate* for action (Lindsay, 1995); or
   (e) it might be based on an assumed rationality presented by science.
   (f) It might be based on common perceptions of science which are not necessarily tenable under scrutiny or agreed on by scientists (see Forsyth, 1996). Nevertheless, such orthodoxies may be promoted in the media and generally believed.
2. Science is not dispassionate nor value free:
   (a) it may refer to its own agendas of *pure discovery and curiosity*; or
   (b) it may be working in a general *applied* context; or
   (c) it may be *problem orientated*; or
   (d) it may be explicitly *policy led*.
   (e) It may attempt to aspire to (a) but be *politically constrained* in terms of what is fundable.
   (f) It may attempt to aspire to (a) but may be *scientifically constrained* by paradigms and what is supported by other scientists through the refereeing system (Lovelock, 1995).

Those who espouse seeking a clear mandate under 1(d) tend to reject the utility of science as pure discovery under 2(a) in terms of not providing useful answers and being critical in terms of scientific constraints under 2(f) and convergent thinking, promoting instead applied, problem-orientated and policy-led science under 2(b), (c) and (d) (Wakeford and Walters, 1995). Those who espouse pure discovery under 2(a) resent the constraints of application and constraint under 2(b)–(f), seeking to foster the spirit of pure enquiry and the ideas of tomorrow, seeing application to present-day problems as short-termism and intellectually undesirable. Interestingly, Boehmer-Christiansen (1994) also writes of the "Limits to scientific advice" and adds that the "need for scientific advice declines rapidly once a problem is actually dealt with by regulatory, technological or behavioural change".

As an example of political priorities 1(a) Colten (1998) writes of policies applied to the American High Plains in the past as "Rooted in popular perception and deep-set values which shaped agricultural development and policy formulations". This was with reference to Opie's (1998) article on *Moral geography in high plains history*, which Opie sees as a "political process that is the outcome of a state of mind" deriving from a situation when "government policy identifies a geographical landscape and its inhabitants in need and deliberately respond to save that region". The deliberate government action intervened to "preserve the family farm as a locus of 'good' human values and 'authentic' environmental conditions" and where agriculture was not seen as just furnishing the country with food and fibre but "a set of children with a work ethic and a good set of values". Science does not enter in this consideration and, indeed, why should it? The necessary relationship between science and policy is perhaps one that is seen by scientists rather than by policy makers. This much is clear from the writings of Anderson *et al.* (1998) in the context of forestry. As the authors point out, there will be a number of competing ideas but it is not a matter of seeing who is "right" with, perhaps insiders or "locals" being "right" rather than outsiders, but in establishing dialogues between all the players. This admits that "there are *several alternative management plans which are consistent with the available scientific evidence*" (emphasis added), and thus that scientific "evidence" does not have the answer for policy – it is how the evidence is interpreted in different social contexts which matters.

Having made such possible differentiations, the further point is that just because these aspects might be thus differentiated, they are not necessarily mutually exclusive. Multiple objectives are perfectly possible and the spirit of pure enquiry might be fostered in applied and policy-led contexts; policy-formulation can equally derive from the results of pure curiosity. Many combinations and clusters of diverse approaches and contexts are possible. What might be at issue is whether any of them are more or less desirable than any others. The friction arises in tackling the questions of whether science is, or indeed should be, useful to policy makers and if policy making is based on science whether this should actually act as a constraint upon scientific activity. Of course the constraint does not necessarily follow from the exhortation for science-led decision making but with limited resources for research, it may act to do so. Simply, if science seeks to be both independent of political influence and also useful to policy making, does the perceived need for the latter constrain the former?

If scientists feel that science should be independent of political influence *and* politicians feel that science should be useful to policy making, is the latter compatible with the former if political direction is to encourage the useful science? The further issue is that the practice of science, with a few notable exceptions (e.g. Bormann and Likens, 1979), is dissective but the environment reacts as a whole: if management should thus be holistic, there are then problems if it is to be based on a science which is dissective.

In considering the relationship between scientific understanding and policy, the following points can be quoted from Trudgill and Richards (1997):

1. Research can often make things more complicated, asking more questions than are answered and often indicating the need to look at local conditions.
2. Policy makers can see such complexity a reason for inaction (which may be interpreted as prudence or, more cynically, as an excuse for not spending money, or both).
3. Policy makers who are willing to act and feel the need for a clear mandate for action from scientists can also find such expressions of complexity exasperating, especially if the policy makers are willing to act on the precautionary principle to prevent future problems.
4. The call for a clear mandate for action can be seen as entirely unreasonable by scientists. This is because things are in fact complicated (or, more cynically, because clarity may imply research project finality and remove the need for further funding or both).
5. Policy making may involve, at best, a simplification of a complex situation. If complexities exist at a local scale, broad scale policies may be wholly inappropriate at specific sites.
6. Policy making decisions may also be undertaken without, or despite, any scientific evidence but for political, economic or social reasons.

Trudgill and Richards (1997) illustrate points 2–4 as follows:

Scientific tensions and uncertainties can be exasperating for those who wish to press ahead or even used as a cover-up reason for inaction. However, it is certain that the wrong action can be expensive and damaging. This can be illustrated by the statement from Nicholas Ridley (the then UK Secretary of State for the Environment) that "it is absolutely essential that before you take expensive and important action you are certain that you know what you are doing. I am very keen to know exactly what it is that the problem is, what the cause of it is and the cure is before trying to take action. One of the most damaging things you can do in environmental policy is to take action on the wrong cause with the wrong cure at great expense without achieving any gain" (quoted from Trudgill, 1989). But can science provide such unambiguous answers?

Tensions might also exist in terms of the complexity of the environment. For example, in 1993 at the third Stockholm Water Symposium the County Governor of Stockholm (Adelsohn, 1994) expressed an exasperation at the inability of scientists to come up with clear and applicable mandates for action concerning decisions about whether or not to build a bridge across the Baltic from Denmark to Sweden. He complained that scientists came up with very complex answers to simple questions about the possible effects on water flow and biota and that the scientific body as a whole often presented conflicting messages to decision makers. The scientists involved felt the overwhelming need for further research on the site before they could give anything like a realistic answer and were roundly criticized by the policy maker. It is perhaps no wonder that some decisions are made on purely political grounds and the arguments then become ones of different ideologies, feasibilities and economics rather concerned with science.

Further constructs as to the limitations of science are evident in realms of prediction. In studying the relationship between environmental change and species response Huntley *et al.* (1997) examined the changes in the Quaternary, asking if the past was a guide to the future. They virtually conclude that prediction is impossible:

Because we remain ignorant of the basic biology of the majority of species, and of their interactions and interdependencies, *the natural world retains a disquieting capacity to surprise us with unpredictable responses* to global climatic change (emphasis added) ... we cannot expect to be able to foresee and predict all the potential consequences for ecosystem function ... Some species whose dispersal ability is artificially enhanced by some form of human activity may, as a result, be able to migrate quickly enough to remain in equilibrium with changing climate (zones). Other species that today are rare may rapidly increase their populations if they are able to exploit the habitats disturbed by human activities and/or as consequence of climate change impacts. Yet other species that today are abundant may rapidly be driven to extinction by novel environmental conditions of the future, perhaps combined with pressures exerted by human population. The extent to which extinction may affect the migratory responses ... will only be revealed when these taxa are called upon to respond to a rapidly changing environment.

The writings, quoted at length above, pose a sense of problem in terms of the inability of science to come up with answers. However, this may not be the case as, rather than saying that ecosystems are mutually reactive, leading them to be inherently unpredictable, science can often provide a "best shot" answer, based on current understanding, with an appropriate expression of uncertainty. This is certainly the more sanguine view of Mitchell and Hulme (1999) who while admitting that climatic change is inherently unpredictable they progress to the specification of the sources of unpredictability in the manner of Lemons (1996). They write that scientific models for climatic forecasting are often seen as deficient. However, they argue that it is not a problem with the models and that they regard prediction as involving "a cascade of uncertainty (which can be specified and involve feedback mechanisms) rather than a single result process sullied by model deficiency".

There may also be a way ahead in terms of identifying preferred states – and of the human capacity to strive for them (which Mitchell and Hulme also recognize as an important consideration). This turns the argument round. Rather than being somewhat dismissive of concepts like "succession" or "maturity" as cultural constructs of nature, cannot we argue that it is actually appropriate to also pay attention to cultural preferences about ecosystems? We might then seek to identify cultural preferences and make our science appropriate to culturally defined goals? This would be more in tune with "ecology for people" (see section 1.1).

In this context, Trudgill and Richards (1997) argued that:

> Funtowicz and Ravetz (1993) refer to traditional core science, applied science and professional consultancy and to a newer "post-normal" science which responds to the challenges of policy issues. A general consideration in the latter is the dialogue and interconnectedness between tackling environmental problems and the nature of environmental science. An older view might have been that environmental scientists undertake their science independently of any application, often seen as fundamental science, and which can then be presented to policy makers, almost to make what they want of it. A more recent view is that not only is science context dependent and not free of values but also that scientists should be addressing more fully the needs of policy makers (Trudgill, 1990, Chapter 4). Here, scientific research is undertaken to answer particular questions posed by the definitions of particular social and environmental goals (Wakeford and Walters, 1995). This latter, policy-led view is in the ascendancy. Additionally, Pacione (1999) now talks of "useful" knowledge.

It can then be argued that:

1. Preferred states can be identified in terms of utility, aesthetics and ethics.
2. Preferred states can be aimed at by manipulation, based on current ecological understanding of the mechanisms involved, with further research if the understanding was felt to be insufficient in the context of reaching the preferred state.
3. The envisaged mechanisms should include that of mutual interactiveness between the components and responsive changes of the components.
4. The element of surprise can only be dealt with by trial and error, learning from experience.
5. A difficulty lies in the fact that while experimentation may aid prediction, experimentation is often reductionist and whole systems (drainage basins, lakes, forests) may react differently than might be indicated by factorial experiments. Large scale experimentation is therefore important (as seen, for example, in watershed treatments, Borman and Likens (1979).
6. What applies to one system in one place may not necessarily be applicable to another place because of the spatially unique assemblages of components and, probably more importantly, the unique histories of places. Calibration of general models and policies is therefore important (Trudgill and Richards, 1997).
7. The past is not necessarily a guide to the future as the starting point is the conditions as they are now, rather than the conditions that existed in the past. An element of uncertainty is therefore inherent in any management action (meaning that a preventative or precautionary approach is preferable – Huntley *et al.* (1995) conclude that it is preferable to curb the emissions of greenhouse gases because it is difficult to predict the consequences if we do not).

This is thus a "goal orientated" approach advocated by Trudgill (1990). Fundamental to this is the notion of hybrid science where we address both cultural/social and scientific issues simultaneously in a specific context. Batterbury *et al.* (1997) write of hybrid research and democratic policy. To paraphrase their understanding, hybrid research uses knowledge claims from a variety of sources, both qualitative and quantitative, and is locally-based (see also Shapin, 1998). It is both social and physical, reliant on indigenous knowledge and experience and allows researchers to be aware of their own social and cultural settings. The democracy aspects legitimizes the views of previously underrepresented groups – the poor and the disenfranchised – as well as the powerful and the knowledgeable. It also helps to overcome the dichotomy between "expert" and "lay" knowledge lamented by Wynne (1996). It also helps in the dialogue between indigenous and scientific knowledge (Agarwal, 1995), dismantling the divide between them which polarizes the situation in that either indigenous knowledge is to be dismissed or is seen as providing the answers (Barraclough and Ghimire, 1995). Thus we might now argue that both environmental perception and scientific approaches are both socially produced and that all attempts to understand externally-real biophysical processes are selective and constructed. The state of the environment is variously constructed from physical and social viewpoints. Essentially, it is different from evidence but stresses the *significance* placed on observations about the environment and its changes (Figure 2.1). The questions of whether changes are seen as significant in terms of being, beneficial or detrimental in immediate terms or of being desirable or undesirable in wider terms are fundamental.

The last word about the role of science in environmental management is one where we ask the question: what are the meanings of the findings? There are three groups of meaning (Figure 2.2).

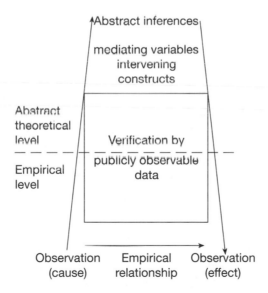

Figure 2.1 Abstract inferences are made about observable phenomena through mediating variables and intervening constructs (from Bell *et al.*, 1996).

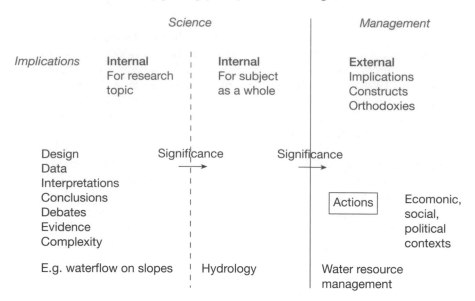

Figure 2.2 Three domains of science – internal to that study topic (e.g. findings from hillslope hydrology research), internal to a subject as a whole (implications for hydrology/rivers) and external – the public domain/management (e.g. quality of drinking water). The information flow from left to right is mediated in terms of the *significance* of the conclusions in terms of meaning and values. In addition, in the public domain, actions are often based on *orthodoxies*, or what is generally believed. Orthodoxies rarely cover all the details, complexities or debates which may be present within the detailed, internal science domain; the more so if public participation is involved, the less so if specialist management personnel are involved.

1. Internal, self-defined meaning. From the data (relative to the experiment/ observation), what conclusions can be made for the topic under discussion. This is often only as far as many scientific reports get in their conclusions. Management may make what it wants of it (or not).
2. Generalized meanings. What are the implications for the whole subject matter in which the research topic is nested (but still internal to science).
3. External cultural/social/management meanings. How may the findings be interpreted in cultural/social contexts? What are the implications for management? How are the findings to be culturally constructed? These are not often explicit unless the research is nested in an "... and management" context.

As a facile example we might move from (1) water is observed to flow downhill to (2) the foot of a slope will be wetter and provide runoff to streams to (3) we should be careful when applying agrochemicals to sloping land as they easily end up in the stream *or* water should be retained on slopes if we are to maximize its utility in agriculture and so on. It is this final realm of meaning which is culturally the most important and the same findings under (1) can obviously be constructed in different ways in differing social contexts. It is the significance placed on scientific information which matters to society and if science is underplayed (in the view of scientists) in decision making then it is surely the scientists' task to attempt to display the possible significances more explicitly in cultural contexts, rather than to bemoan the fact that their science is not being used.

Scientists tend to retain their putatively "objective" stance without realizing two groups of things. First that, as Darier (1999a, b) writes in *Discourses of the Environment* that Foucault's questions were "*How have my questions been produced?*" and "*How has my path of knowledge been determined?*" together with "*How have I been situated to experience the real?*" (emphasis added). Thus "why am I asking this question, why am I going about my research in this way and have these reasons and methods influenced my findings?" are rarely asked by many scientists. Posing these questions may ease the transition from (1) to (3) above. Second, in trying to retain "objectivity" scientists rarely like to explain the significance of their work to society in a precise context. This is because it places it in a context which is value laden when they might have made a conscious attempt to divorce their work from specific "applied" contexts. Given that Foucault's questions mean that their work is not context free, it might be as well to attempt to be explicitly conscious of the contexts in which they are implicitly and unconsciously working – to examine basic (unexamined) assumptions – and to be then explicit about the wider implications in the possible range of value-laden contexts. Scientists will, of course, argue that their work might be significant, and/or used in a number of contexts, but if we are to be guided by science in any way, for each piece of scientific research that we read, we can specify a context and then we should ask: "OK, but *what is the significance of your finding?*" Hopefully, then, we are able to enter into a dialogue with the researcher to check the basis of decision making where the science may be used in a specified context. The question is whether the scientist does, or contributes to, the translation to the policy context or whether it is left for policy makers to make what they will of the science. A dialogue between scientists and policy makers does appear useful in that context dependency is crucial: science is both produced and used in specific contexts and it is these contexts which give the science significance and meaning.

Science as a "continual process of criticism" (Dunbar, 1995) and as "concerned with the unknown" (Lindsay, 1995) is highly defensible, but hardly helpful to a management endeavour which has concerns about taking action on the basis the body of "what is known" (see Trudgill *et al.* (1999) *Water Quality: Processes and policy*). If the knowledge is needed in applied contexts, it would be better if it were explicit that, and how, such a "body of knowledge" is both culturally derived and employed. We would then not be deceived by the concreteness of science (see Rawles (1998) on *The fallacy of misplaced concreteness*) but take it as provisional guidance – science's "best shot" so far.

2.2 Can we be guided by "ecological principles"?

If we follow Simmons (1993a) in Sheail (1995) the simple answer is "No": "As Simmons (1993a) warned, the reading of moral and ethical tenets from the natural world not only ran the danger of simply projecting onto nature what the author wanted to believe anyway, but it (the reading) took no cognizance of the misgivings and debate within ecology as to the applicability of such concepts, even in the natural world." We quoted this in Chapter 1, implying there that the derivation of "ecological principles" from nature was tautological since nature was a social construct so we were merely deriving our guides to management action from our own values. While this point may be philosophically valid, it is not, however, a negative one. What it does mean is that there is a direct linkage between our value systems, our views of nature and our proposed actions. Rather than trying to derive "principles from nature", then, we can say that the answer to the question "can we be guided by ecological principles?" could be, in fact, "Yes" provided we realize that the "principles" are human constructs. In essence, it is more pertinent to ask: "What, then are the relationships between conceptual models, constructs of nature, attitudes to the environment and management policies?" Might we go straight from value laden constructs of nature to value laden policies without the intermediate stage of projecting our values onto nature and then deriving illusory principles from nature as constructed?

A number of authors have addressed this question, either with or without the intervening step of seeing nature as a social construct. Shrader-Frechette and McCoy (1993) argue that "*ecology does not and will not yield general and deterministic laws useful to environmental management*". Watt (1973) felt that "nature is immensely complex, often in ways about which our ignorance is awesome. Consequently, when we make major changes in the world, deliberately or, as is more frequently the case, accidentally, we are modifying enormously complex homeostatic mechanisms in such a way that *we may not be able to foresee the consequences. This is why surprising things sometime happen*" (emphasis added). Uncertainty in managing ecosystems is stressed by Carpenter (1996) though there is the feeling that the type of errors which can be made might be specified. If the null hypothesis is that human intervention causes no effect on an ecosystem, then a Type I error is to reject a null hypothesis which is true (concluding that a change has occurred when it hasn't) while a Type II error is to accept a false null hypothesis (concluding that there has been no change when there has). None of this, however, is again particularly helpful in the context of the wishes of environmental managers wishing to predict the outcome of their actions or inactions.

If we wish to seek a way forward, we can say that so far we have seen that the study of the terrestrial biosphere involves investigating the relationships between land (and the vegetation or crops growing on it together with the animals they support), the wider environment and people using the land. This involves the study of:

- Land resources (amount, quality and availability/distribution).
- Vegetation resources (amount, quality and availability/distribution).
- Utilization of these resources – and questions of social equity during this utilization.

- Environmental impacts of this utilization, and spatial variations in these impacts.
- A discussion of social and ecological priorities.

On the latter point, if we are able to identify ecological principles and priorities we can see that we could devise management plans based on ecological priorities (species diversity, habitats, self-renewal and so on as shown in the quotation from Barbier (1987), above (p. 37), as biological goals). This can be seen as a science-led approach. Equally, we can see that we can devise management plans based on social priorities (well being, food security, shelter). This is where social goals dominate. If they do not lead to the same management plans, we should then ask, why not? Additionally, if successful environmental management is to be based on an adequate knowledge base, what kinds of knowledge do we need to have and how certain can we be about the knowledge?

Certainly, we should seek to be able to predict the ecosystem response to any proposed management action or external environmental change, together with any specification of uncertainty. As a consequence, we should seek to understand how ecosystems work in order to be able to do this. It is also commonly seen that such an understanding is a fundamental prerequisite in that it is easier and, in principle, cheaper to work with nature than against it (see Jordan (1998) in *Working with Nature*).

One fundamental element of our understanding is that of linkage and interaction. We know, partly through observation and partly through experience of the consequence of management actions that a single component is unlikely to be related to just one other component and that single action is unlikely to have a single consequence. We should therefore expect that there might be multiple consequences deriving from any one action.

A second element, and one that devolves from the first, is that of flow, storage and cycling of energy and matter. In this context, there is nothing stable within ecosystems – matter and energy that is present within one site or life form is liable to be within a different site or life form over time. The time scales vary but we must expect that something which is in one place at one time will in all probability be in one place at another.

A third element, which also devolves from the first, is one of adjustment and reaction. Any ecosystem will have evolved in relation to its external conditions (and may have even influenced them). The adjustment is both to external conditions like climate and internally with respect to the components of that system adjusting to each other. We must expect then that both external and internal change will be followed by adjustment reactions, particularly in relation to thresholds, the relative importance of the components and the ways in which they operate.

A further point is that the functions which species perform can be seen as more important than species themselves. Ecosystems are functional in that there are groups of individual species which will perform similar functions – the decomposers, the predators and so on. In this sense species do not matter as long as the function is being performed – without decomposers, the ecosystem would become full of waste matter; what matters is that the *type of organism* is there rather than whether they are *particular species of all the organisms that perform the functions* of decomposition at its various stages, like flies, bacteria and earthworms. *The components are self-defining components because they both make up the ecosystem and permit it to exist.* In this sense biodiversity, while it might be a human preference, does not matter but ecosystem function does. In this context, biodiversity can be seen as important but only in the sense that each ecosystem function should be fulfilled. However, this is not to say that every ecosystem must have each function fulfilled or each function fulfilled equally. Where decomposition is limited, waste products do simply accumulate – as is the case with peat bogs. Where predators are absent, numbers become limited by food supply rather than by predator. The relative importance of the components merely reflects the degree of importance of the functions.

The key question then becomes one of how we link functions with values. In a book on *Ecological concepts: the contribution of ecology to an understanding of the natural world* (Cherrett, 1989b), the chapter titles reveal a view of the organizing concepts:

- Fluxes of matter and energy.
- Food webs.
- Ecological niches.
- Diversity and stability.
- Predator–prey and host–pathogen interactions.
- Population regulation.
- Competition.
- Life-history strategies.
- Optimization levels of organization.

In a survey of the importance of concepts "the ecosystem" was seen as the most important concept followed by the next top ten of succession, energy flow, conservation of resources, competition, niche, materials cycling, the community, life-history strategies, ecosystem fragility and food webs, in that order. These are, however, the concerns of ecological scientists. If we attempt to identify the concepts which might be of interest to most people (see Harrison and Burgess (1994) and also Irwin (1995)) they could be seen as first coming from our concerns:

Human concerns about ecosystems

- Provision of food and shelter from ecosystems
- Methods of production from ecosystems
- Equity of distribution of ecosystem goods
- Market – economic viability of ecosystem management
- Meaning – spirituality, values, security and recreation from ecosystems.

This might give us:

Ecosystem concepts of interest to people

- Productivity of the ecosystem
- Renewability of the ecosystem
- Inheritance from the past
- Sensitivity to change.

2.3 Ecological principles as value judgements

There is a long history of the notion that science is the antitheses of the development of values. In his book *Philosophy for Our Times,* C.E.M. Joad (1940) wrote about the position that science can give neither a satisfactory account of mind nor of values. He saw science as only dealing with the measurable aspects of phenomena: "Science, I have argued, can only deal with what can be empirically observed and what is mechanically caused". He turns to examples from music, poetry and art and writes how science would, for example, see music as the relationship between neural impulses and a series of black marks on paper and, on art, he says that *all* science has to tell us is about the chemical composition of the pigmentation. This may now be seen as somewhat unfair to science, but the lasting point is that there are systems of values and feelings which are the most important. In terms relevant to ecosystem science, he precedes Golley (1993) by saying how "science takes whole things to pieces and describes them in terms of pieces" and writes of the importance of "wholes which are more than the sum of the parts" and that value stems from integrity. He then discusses the possibility that values are figments but proceeds to argue that values are, in fact, real and indeed "ultimate". These are fascinating writings for our times as well as in 1940 and highly relevant today when we might argue firstly that eco systems are more than just a social construct and integrity is seen as valued. There is, however, something further than that, which is a statement about desirability.

Human beings make value judgements about ecosystems, both in a scientific sense and a spiritual sense. Scientists try to relate productivity to diversity, diversity to stability but in truth they are attached to the notions of stability and diversity as perceptions of ecosystems. Ecosystems can be productive and diverse, productive and monospecific; stable over time and diverse or monospecific; unstable over time and species rich or species poor. Human beings have a heritage of being insecure within the environment and its hostility and seek security through food security and shelter. They also seek the security of knowing that something is there when they return to it after a period of time has elapsed. This is intrinsic of much of nature conservation, a great deal of which is aimed at arresting change in vegetation (see Chapter 5). It is no more than a human psychological need for reassurance which manifests itself as a scientific notion of conservation management. Similarly, the notion of diversity is seen as markedly important but is no more than (and no less than) a human sentiment which devolves from awe, wonder and a feeling of stewardship: the ecosystem does not need to be diverse. The notion that ecosystems are more stable if they are diverse (because if one species is made extinct another can take its place – as palaeontology shows) is again a reflection of the human desire for order and a sentimental attachment to some animals and plants which are regarded as charismatic. Ecosystems, if they exist at all, do not need to be diverse or stable and while any necessity is our construct, they are more fundamentally constructed as merely multi-componented, inter-reactive, exhibiting flows of material and energy, changing over time and reacting and adjusting internally to the varied components and to external conditions, with their visible forms being merely manifestations of the processes operating. It is human value judgements that cause us to talk of stability and diversity and it is the spiritual, cultural and utilitarian values we attach to ecosystems which makes us indicate preferred states and use such words as "pristine" and "untouched" and our spiritual, aesthetic or utilitarian response which makes us value one species rather than another. Simply think of the notion of a weed, or the preference for a gorilla or elephant rather than a centipede or worm. Entrenched in all this is the notion of wilderness, biblical notions of the "garden of Eden" and other constructs which colour our view of ecosystems.

So when we talk of deriving ecological principles, we have the notions of inter-reaction, flows and adjustments which as far as is humanly possible are not completely value laden. These are separable from those concepts which reflect desirability which reflect human spiritual response, aesthetics, cultural values and utility.

So if we think about basing environmental management on ecological principles, all we have are the foundational concepts that ecosystems exhibit flows and their many components inter-react (see Inset 2.1). The rest of the time we are talking about preferred states or species, preferred landscapes and preferred uses together with the perception of, and attachment to concepts of "naturalness". It is in defence of preservation that we use the concept of naturalness which is as much about guilt about past destruction and the inner need for something higher than ourselves than it is about the "need" to preserve something that is "natural". In part, also, we can suggest that it is a clash of preferences which lead to the perceptions that environmental problems exist and not necessarily that ecological principles have been flouted. Ecosystems rarely "collapse" but change from a preferable to a less preferable state. Imagine a field where the soil was full of pests but these were being eaten by an attractive bird, (say, in the UK, the lapwing): we might talk of the balance of nature being upset if someone shot these attractive birds and they were replaced by the management introduction of less attractive birds (like the UK starlings). The latter might eat just as many pests and the balance between predator and prey would be the same and the ecosystem would function in the same way but it would (a) not be like it used to be and (b) have less preferable species. We would probably say that biodiversity had suffered because of the loss of the earlier

Inset 2.1 Nutrient cycling in ecosystems

Nutrient cycling in ecosystems (from Trudgill, 1988). Does the pathway B–F–H–S–B provide a guide for human society? Such recycling of nutrients can be seen as a mandate for a political ecology based on ecosystem science. However, note also that there are pathways M–S, S–L and A–S: nature exploits resources irreversibly in M–S and exports material irreversibly from the ecosystem S–O as well as passively receiving inputs in A–S. The claim that ecosystem science provides justification for the promotion of recycling in society is thus not without basis but can also be seen as flawed.

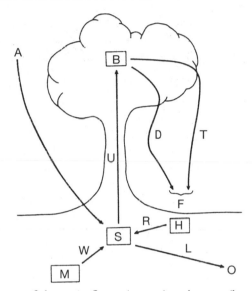

Simplified diagram of the main fluxes (arrows) and stores (boxes) in a forested ecosystem.

The fluxes of nutrients are indicated by arrows, and stores are indicated by boxes, both with shorthand notations as follows:

Fluxes:

A = Atmospheric inputs – wet and dry deposition.
W = Weathering inputs from bedrock and soil minerals.
U = Uptake by plant roots.
D = Dead biomass fall, litter, stems, and tree trunks.
T = Throughfall and stemflow.
F = Forest floor inputs (D + T).
R = Return of nutrients from vegetation via humus to soil store.
L = Leaching through the soil.
O = Output to runoff waters.

Stores:

M = Minerals in bedrock and soil.
B = Biomass storage – roots, stems, and leaves.
H = Humus store, nutrients in dead biomass not yet released by decomposition.
S = Soil store of available nutrients – soil solution and absorbed on clays and humus.

species. Ecological principles (in respect of function) would not have been flouted but preferences would have been offended.

Ecological principles alone cannot then provide a mandate for desirable actions, even if we can be clear about what those principles might be. They can guide us to the likely general outcome of the consequences of our actions – we are likely to see consequences wider than any specific management target. However, they cannot indicate the "right" option out of several possibilities; these must be the province of human preferences. Everything will always be recycled – whether it be sewage taken to the land, water or sea – and water chemistry and species composition will change accordingly. It is a human notion of preference and of ethics as to whether it is right to recycle material back to source. Ecosystems in general, but not exclusively, recycle internally but it is a human judgement that this is preferable. Ecological principles are thus a human value construct placed upon the operations of nature which people perceive. As such, they are highly charged with the notion that something is better if it is natural. There is a human satisfaction in organic farming in that we are being natural and recycling material and also that we feel in tune with nature. There is a moral higher ground. We are using the "ecological principle" of recycling but it is, in fact, more about feeling that you are not doing any damage and illustrates that "ecological principles" are a value-laden construct rather than something somehow objective. To illustrate this, we might usefully look at the concept of an ecosystem as proposed by Tansley (usefully reviewed by Sheail, 1987) as a foundation for actions but then realize that the concept is itself value laden (see Chapter 3 and Cameron, 1999; Cameron and Forrester, 2000). We are here dealing with the realm of social priorities.

Perhaps a more interesting question is: how do we derive ecological priorities? Here, we appear to be acting on behalf of the environment and somehow deriving a "higher" principle from deductions about the workings of nature. In this context, we may also ask again (see also p. 17): "Are social priorities different from ecological ones?" If they differ, then this surely must be the cause of environmental problems as perceived by at least some people? While we must realize that an environmental problem is a social construct and, almost by definition, involves issues which may be constructed by some people as a problem and not by others, if environmental damage is perceived we should ask: "*Why do we damage the environment?*" Damage, itself, is a social construct and subject to discussions, disagreements and various definitions but we can begin by suggesting that there must be a difference between social priorities and ecological priorities (while simultaneously recognizing that both are difficult to define). Earlier, we generalized about environmental issues but now it would be more appropriate to take a more detailed view. We can think of several reasons for situations existing where we might have the perception of damage (however constructed) to the environment:

- Short-term views are held rather than long-term ones.
- Priorities other than the environment dominate.
- Power and/or war can be causing damage.

- Territorial imperative and nationalism may dominate.
- Economic rewards/the cheapest option dominate.
- There are inadvertent environmental consequences of other actions.
- Greater food production leads to unsustainable development.
- Immediate needs for shelter and fuel dominate.
- There is diffuseness of the causes of problems which make them difficult to tackle.
- Lack of awareness of effects of actions.

While all these deserve to be defined and expanded upon, they may be resolved into some categories:

1. *Significance*. The effect was dismissed as not being significant.
2. *Knowledge*. Lack of awareness of effects – scientific knowledge inadequate/uncertain or adequate but not transmitted.
3. *Economics*. Damaging effect was cheaper, more remunerative/or easier.
4. *Distribution*. Distribution of the effects was disparate between the effector and the receptor so that the benefits/damage were unequally distributed amongst people.
5. *Incompetence*. Inappropriate individual actions/managerial structure.
6. *Priorities*. The environment was not a priority.

Which basically come down to:

Information, Cost and Social structure

Simply, if we propose elements of failure, they are ignorance of the effects (not foreseen or not known or, alternatively, known but unavailable to practitioners or regarded as not significant); the costs were not covered by the benefits and the social organization was such that safeguards were not in place, the costs and/or benefits were unequally distributed and/or other priorities dominated.

If we propose elements of success, they are appropriate information held by the enactors, the costs were covered by the financial benefits and all those involved or implicated benefited.

Our challenge is thus to ensure that future management is well informed, economically viable and socially appropriate. Our questions become:

1. What are the environmental changes involved?
2. How does the biosphere react to them?
3. Are the changes thus seen as detrimental?
4. If so, can we do anything about them?
5. Should we actually take action?

The agenda for tackling them then becomes one of both social and physical knowledge. We need:

1. An appropriate understanding of the physical environment.
2. A realistic statement about current conditions.
3. A value judgement as to an ideal state, referring to both physical and social conditions.

4. An appropriate understanding of the social contexts so that we can be practical about our idealism.

Finally, however, we have to decide on our social constructs before we can say what is appropriate and what is a desirable state that we want. This is a topic to which we shall return in Chapter 9. For the moment we may again cite the work of Opie (1998) who firstly saw agricultural policy for the American Great Plains as driven by the value systems in people's mind. Secondly, Opie showed how there were at least six constructs of the Great Plains which might motivate policy. Opie concludes that the Great Plains might be seen as:

1. A geographical zone of good soil and dry climate.
2. A manifestation of US Destiny, including triumphalism.
3. A breadbasket which was highly productive.
4. A growing pile of debts to the bank.
5. A cherished and valuable lifestyle of the yeoman farmer.
6. Similar to other areas.

The list could be added to but this perceptive paper realizes that the way the area was mentally constructed had a great influence on policy. Science has a great deal to contribute in the simpler, mechanical, verifiable systems, but in the "loose", less tangible and unverifiable constructions like ecosystems and landscape *science is a foundation but it has little to offer in the sense that it can't decide what views people will hold and how they will act.* For our purposes, we can say for the moment that the way ecosystems are socially constructed has a strong influence on our policies and that ecosystem science is part of that construction. It leaves the question of "how should we construct concepts of ecosystems?" for the next chapter.

Further reading

> ** Key reading
> * Important reading
>   Recommended reading

Drury, W.H. (1998) *Chance and Change: Ecology for conservationists.* University of California Press.

Dunbar, R. (1995) *The Trouble with Science.* Faber and Faber.

Eden, S. (1996) Public participation in environmental policy: considering scientific, counter scientific and non-scientific contributions. *Public Understanding of Science,* 5, 183–204.

Harrison, C.M. and Burgess, J. (1994) Social constructions of nature: A case study of conflicts over the development of Rainham Marshes. *Transactions of the Institute of British Geographers,* 19, 291–310.

Irwin, A. (1995) *Citizen Science: A study of people, expertise and sustainable development.* Routledge.

*   Jordan, C.F. (1998) *Working with Nature: Resource management and sustainability.*
     Harwood Academic Publishers.
**   Lemons, J. (1996) *Scientific uncertainty and environmental problem solving.*
     Blackwell.
*   Trudgill, S.T. and Richards, K.S. (1997) Environmental science and policy:
     generalizations and context sensitivity. *Transactions of the Institute of British
     Geographers*, 22, 5–12.
*   Wakeford, T. and Walters, M. (1995) *Science for the Earth – can science make the
     world a better place?* Wiley.

Chapter 3

Ecosystems: the conceptual basis

Summary of key points:

1. "Ecosystem" is a mental construct and should be seen as such.
2. This does not detract from its utility as a concept, simply that its limitations should be realized.
3. As well as interconnectedness, there are two important aspects: heritage and recycling.
4. Ecosystems are renewable but not repeatable.
5. Change is evident; stability and other value judgements are placed on these changes.
6. The key point of the ecosystem concept is that we realize that if we do something it may have consequences beyond what was intended or immediately tangible.

3.1 What use is the ecosystem concept?

... the most profound joy for any scientist was when the abstract workings of the mind found a correspondence in nature, in the world we live in.

William Boyd (1991) *Brazzaville Beach*, Penguin Books

There is a flourishing science, Ecology, which deals with the relationships of plants to their environment and one another. For plants, it has been discovered, are not scattered so indiscriminately as they seem to be, but occur in regular societies, composed, as a rule, of a very few dominant species and a number of lesser ones. Each of these societies is restricted to a certain type of soil, and where the soil changed a new set of plants comes to the fore.

Nicholson, E.M. (1929) *How Birds Live*. Williams and Norgate

We relate to the world around us through a series of mental constructs. These form our perceptions which give meaning and significance to experiences and guide our behaviour. The constructs derive from other people through our upbringing and training, and progressively as we become ourselves and not just our parent's child, from reading and the visual media and reflexively from experiencing the outcome of our actions. Constructs of meaning have elements of individuality, because each individual has an individual genetic make-up and has had a unique history, but they also have elements of communality because societies are normative or at least are made up of normative groupings in co-existence. An analogy of how constructs about the same thing can vary while having an element of communality is how a range of different daily newspapers will all cover the previous day's news, there is great variety in emphasis between and within the stories covered, so much so that it can almost seem that there are a range of different worlds being portrayed. However, there will be common themes as to what constitutes news and underlying norms to any views expressed. Portrayal and depiction thus can tell you as much about the portrayer (and any intended audience) as much as about the events themselves.

The purpose of discussing this is that when we are discussing the environment, we must indeed not necessarily expect any consensus, even in terms of self-interest; rather a range of varying view points and perceptions with differing opinions and preferences in relation to unique personal histories. However, there may also be commonly held views present, especially within different groups of people. Tensions may obviously arise if environmental management involves assessing collective views and/or the views of different groups. The critical factor is the way a society is organized, how and where views and knowledge are held and positioned and how these relate to effective power in the society.

I once stood with a group of overseas students from a wide range of countries overlooking a view (Plate 3.1) of grass fields with cows, hedges and woods. I asked the group what they could see and how they felt about it. A range of views came back from "delightful", "productive", "English", "rural" to "threatening", "open", "bleak" and so on, I thought in all probability relating to their backgrounds and experiences, indeed very much so when one said "I thought we had come to look at agriculture, but there isn't any" – there were no ploughed fields

Plate **3.1** The beholding eye: different respondents described this view as "delightful", "productive", "English", "rural" to "threatening", "open", "bleak".

visible and these were what agriculture meant to her in her country where cattle were left to graze around the arable fields rather than having a whole grass field to themselves. So they were perceiving very different views and arranging the world in their minds, including this "new" world I had bought them to; they could then fit this new experience into their existing constructs or adopt (or challenge or modify) a new one I could give them by me saying what I saw and felt about it. I talked of a physical environment which had evolved through geological time, with attendant organisms, and which had been arranged through more recent history according to collective and individual human activity in relation to economic, social and political factors and in which a number of organisms had been encouraged or allowed to grow according to their utilitarian and spiritual value. The countryside indeed had been physically arranged according to a history of mental constructs acting upon physical and biological material which had occurred at this location through time. So I gave them my construct of a constructed environment.

Thus we arrange our cultural perceptions of the environment in two ways – in our minds and physically, but in both we are influenced by what others have told us and what we have learnt through experience. The physical environment, that is, the world other than the self, tends to vary in the degree to which it has been arranged by human activity according to collective and individual mental constructs. This is not without some degree of reciprocity in that the way the world appears to have reacted during this arrangement will have modified our mental constructs, though again this will be through the filter of perceived significance. In this relationship between mental constructs and physical experience, we tend to formulate guiding concepts which might seem to serve us well, though they might also serve to constrain our outlooks. Studies of the history of ecological thought (Worster, 1994; Golley, 1993) show how these concepts have changed over time – and as much through the varied influences upon ecological thinkers as through experience. As a result, we now have various terms which are contested to a greater or lesser degree such as "ecosystem", "disturbance", "recovery", "succession", "community" and so on which are used as mental constructs to internalize and make sense of the world (indeed, in 1989, members of the British Ecological Society thought "The Ecosystem" was the most important concept in ecology, Cherrett (1989a)).

We should, however, see these concepts for what they are – mental ways of organizing the world around us – and not testable or verifiable in the sense that the word "hot" can soon be testable and verifiable by a child at a very early age. Tansley (1935) who developed the concept of the ecosystem saw ecosystems as "mental isolates" for "the purposes of study" (quoted from Rebele (1994)). Ecosystems thus only exist if we think they do. The point of making this point is one of realizing our limitations at any one time in the context of the posit that *the way we view the world very much influences how we deal with it*. What is then interesting is that *our experience may influence our ideas*. Indeed, Darier (1999b, p. 224) cites Foucault's questions as "How have my questions been produced?", "How has my path of knowledge been determined?" and "How have I been situated to experience the real?" For Tansley, according to Cameron (1999) and Cameron and Forrester (in press), believed very much in disturbance and recovery in the human psychoanalytical field, and indeed worked with the psychologist Sigmund Freud. It seems then unsurprising that these notions should become embedded in Tansley's ideas about ecosystems and that, derivatively, the notion of the "health" of ecosystems should also become part of the concept of ecosystems (Dewberry, 1996). What is also interesting in the subsequent questioning of the ecosystem concept (e.g., Wiens (1984) writing of "conceptual issues and evidence", Pomeroy and Alberts (1986) on "Concepts of ecosystems" and Pahl-Wostl (1995) on "Chaos and order") is whether we change our ideas in the light of experience (reversing the posit) or by a more abstract evolution of thought. When one holds what might be termed a "loose" concept like "ecosystem" (as opposed to the immediately verifiable "hot") it might be possible to gather information to support a range of possibly contradictory abstract concepts. It is tempting to say that there has been greater progress in the evolution of ideas about the environment than in testing them. This is possibly because many are untestable or that evidence can always be found to support anything! Indeed, Pimm (1991, pp. 7–9) makes very much that point, writing that the assertion that "simple ecosystems are less stable than complex ones" held sway with very little justification or illustration. Pimm also writes (p. 388) that "if you look at random data long enough, all kinds of patterns can be imagined" and "community ecology is so conceptually complex that we rarely have any difficulty in explaining any result we encounter. Theories tell us where to look and when we readily find what we are looking for, we gain confidence in both the theories ... and in the data". So what use then is the ecosystem concept?

A point of the ecosystem concept is basically the notion of a complex in which the parts interact to produce the behaviour of a whole, with the attendant notion which requires the wholes to have genuine properties – those which are unique to the whole and not reducible just to the properties of the components (Brennan, 1988; Golley, 1993) (see Plate 3.2).

Shrader-Frechette and McCoy (1993) feel that the ecosystem concept has limited utility to management but Golley feels that the ecosystem concept has been useful in environmental ethics, human ecology, ecological economics and in biological conservation but this, I think, is largely through the general idea of interconnectedness – and the pause for thought in that *if we do something it may have consequences beyond what was intended and is immediately tangible*. In this sense the ecosystem concept is an important contribution to building mental constructs because if we are arranging things in our heads and/or the world about us there is now a further element of *consequences beyond experience*. The other important contribution is one of mutual reactivity – there are rarely single consequences, but a number of consequences and these themselves can provoke further actions mechanistically, synergistically or symbiotically, though this awareness does not help much with creating predictability, indeed it tends to decrease it. However, we do find it currently useful to think of ecosystems as possessing the property of mutual reactivity. This has become as much a matter of ethics as one of science (which still often remains dissective) as *the ecosystem*

Plate 3.2 Slapton Wood, Devon. It doesn't have to be seen as an ecosystem – but we can construct it as such. Does it react as a whole? Is it chaos or system? Is it a complex in which the parts interact to produce the behaviour of a whole, with the attendant notion which requires the wholes to have genuine properties – those which are unique to the whole and not reducible just to the properties of the components (Brennan, 1988; Golley, 1993)? Is it just a collection of trees, plants and animals which have evolved over time with some human influence?

concept imparts a sense of moral responsibility to think about the wider consequences of what one is doing rather than just the immediate, the obvious or the intended.

This realization now seems to guide much environmental management thinking – we look for effects other than the target effect (lodged in most people's minds as "pesticides kill birds as well as pests" and so much so that it is difficult to find a tract that doesn't cite the 1962 Carson Silent Spring book) and this mental construct will I think continue to be the greatest legacy of the history of ecosystem thinking and what makes it useful. There are many different contested views about ecosystems, their value and nature (Cherrett, 1989b; Pimm, 1991; Golley, 1993; Walker and Steffen, 1996; Worster, 1994) but the utility of the ecosystem concept lies in the attendant *consciousness of interconnectedness*. Indeed such a concept of interconnectedness was the basis for the book by van Dyne (1969) on the "ecosystem concept in environmental management".

In this book we are looking at the nature of environmental change and ecosystem response. This assumes that there is environmental change and that ecosystem response will be functionally linked to it. This is a great test of our knowledge because we are thereby forced to look at what we think we know about the nature of the functional links. We should therefore examine the mental constructs of functional links and of interconnectedness.

Some strides have been made along this road by VEMAP Members (1995), who have extrapolated existing knowledge of the links of photosynthesis with carbon dioxide and temperature but still ended with an air of uncertainty; Huntley *et al.* (1997) concluded (p. 504) that the lesson of examining the past was that we were so unsure of the consequences of environmental change, we should make every endeavour to stop global warming. In addition, the past may well not be a good guide to the future, even if we think we understand the past.

What is really interesting is that environmental change is a test of the ecosystem construct: what exactly is linked to what? How exactly will things alter if something else changes? Does the whole have genuine properties – those which are unique to the whole and not reducible just to the properties of the components? If the components change, will the whole change too and how would we tell? The reality of global warming is contested, as many things are, and its contested nature is outside the scope of this book, but the challenge to our constructs of interconnectedness is fundamental and therefore central to this book. We indeed use the ecosystem concept as a construct for the book because the understanding of interconnectedness is both a core of the ecosystem concept and basic to the current concerns about environmental change. And since we admit that ecosystems are cultural constructs about the environment (and which refer to the relationships between our mental constructs and our physical arrangement of the environment), we can do no more than structure the discussion in terms of the degree to which we have arranged our environment – from the constructs and physical natures of wilderness through to domesticity. Simply put, if the aim of this book is to ask the question of how do ecosystems respond to environmental change we need to examine our knowledge of how we think ecosystems relate to the environment.

I look out of my study window and see a cottage garden with holly tree and lawn, with flower beds and house martins flying around under the eaves in the sun. How will this all change with global warming if it really is happening? Will the lawn shrivel or will I or the next owner adapt and use more water, or will there be enough water to go round? Might it not just be the same only with the trees more mature? Or will new species thrive? Who knows what cultural adaptations and ingenious responses there might be. One thing is certain, it is interesting to think about human responses to change and the balance between accepting changes and changing things yourself. One thing I do know is that I derive pleasure from looking at those things I have done myself in the garden, perhaps it makes me feel more secure and less ephemeral – people do seem a lot better at changing things and making their mark rather than leaving things alone.

3.2 Renewable but unrepeatable

If we are attempting to have an appropriate understanding of ecosystem processes, a fundamental point to start with is that ecosystems are renewable but unrepeatable. Two important concepts are involved.

1. Recycling.
2. Heritage.

Within ecosystems there are flows of materials (nutrients, water) and energy and associated stores and transformations of biomass. These flows are infinitely sustainable through the processes of photosynthesis producing plant biomass, the ingestion of this biomass by animals and the decomposition of biomass which returns the materials and energy for re-use. The key element is the one of *return for re-use*. In this sense, through the life processes, everything can be renewed.

Systems, however, evolve through time and thus they also have a history. This cannot be repeated. The actual state of any system at any one time relates to the temporal sequences it has gone through. Over time, system components can be constantly broken down and re-formed, but there is also an element of storage, or "memory" in the system – often a storage of the material consequences of past processes, and of materials often valued.

Conservation thus means two things – protection of the processes involved in renewal and protection of valuable stores from past processes (this is separate from nature conservation which is often a protection of states which we have grown to like or prefer, as discussed in Chapter 5). Simply then, environmental management which is undesirable damages the renewal processes (e.g. damaging the growing points of grasses; removal of seeds from a site) and decreases stocks of irreplaceable material (e.g. soil erosion). What is desirable is the protection, or even an enhancement, of the renewal processes and the protection of the irreplaceable.

Thus biomass productivity and decomposition are fundamental processes which we need to understand together with their relationship with irreplaceable components. It is also important to stress that there may be (a) a functional relationship between the irreplaceable material and the renewal processes and also (b) a spiritual feeling that it is unworthy to destroy or damage an ecosystem that has evolved over time in a unique way.

3.3 Conceptual models, science and attitudes to ecosystems

The way we view ecosystems has a strong influence on the way we treat them. Simply stated, if we think of them as fragile and irreplaceable we might be more careful in our management of them than if we think they are robust and resilient. Clearly, there will be a spectrum of views and orthodoxies but in these contexts, the writings of Wiman (1991) and Holling (1986) are useful here. In considering resilience, they suggest that (a) there is the view that nature is benign, self-adjusting and resilient, returning to prior states after disturbance and (b) there is also the view that nature is linear, with continuous progressive change in the face of disturbance. These two views are not, however, mutually exclusive and elements of both may be present, depending on the presence of a mechanism for recovery. Organisms have different adaptations to different conditions and therefore changing conditions may lead to changes in frequency of occurrence of species: a species which may be marginal under one set of conditions may dominate or die out under changing conditions. The notion that an ecosystem can recover after disturbance is a well established one and is discussed as one of the unifying concepts in ecology by van Dobben and Lowe-McConnell (1975). The concept has both a functional basis and a psychological appeal involving the human perception of a desired state, upon which a definition of recovery might be based.

It has been argued that increased species diversity facilitates recovery and stability, though as Orians (1975) argued, diversity and stability are not necessarily linked and species diversity may well be a reflection of other factors, for example of marginal or limited environmental conditions. In nutrient-rich environments, for example, one or two species may become dominant through being able to out-compete other species while greater diversity might be allowed in nutrient-poor environments because nutrient limitation will limit the growth of potentially dominant species. This does not mean that either situations is potentially more stable (in terms of the continued presence of existing species) than the other. Functionally there is only the mechanisms of adaptation to environmental conditions (including disturbance) and the competition between species which gives rise to the assemblage of species currently existing. This assemblage may then well change over time if the growth of plants themselves induces changes which makes the situation more suitable to the growth of different plants possible and/or allows the growth of slower-growing plants progress. "Recovery" is more a human judgement of the value of the state prior to disturbance. Involved here is the notion of the "balance of nature" which can be "disturbed". In terms of inputs, outputs and stores of water, energy and matter, nature is always balanced by associated adjustments of functions (for example, organic matter is either decomposed and the components recycled or stored in an accumulating humus layer if decomposition is inhibited). Destroy all the predators in a system and the prey multiply till they are limited in numbers by food supply. The concept of the "balance of nature" has little to do with functional balance and much to do with preferred states and the notion of "pristine" conditions "disturbed" by human activity (see Chapter 4).

Much of this discussion relates to the narrative that mankind has been the greatest disturber of all and before such disturbances nature was balanced. There exists a notion that current changes, largely seen as anthropogenic, are more extensive than have ever occurred before. However, as we said in Chapter 1, one only has to look at the geological record, and indeed evidence from the Quaternary to see how climatic changes and glaciations have had a huge impact on the landscape and ecosystems in every conceivable way and to a massive extent. The book *Past and Future Rapid Environmental Changes: The spatial and evolutionary responses of terrestrial biota* (Huntley et al., 1997) shows that the lessons from Quaternary environmental change involved both changing geographic location and also elements of adapting to new conditions: "Few species are likely to have survived Quaternary climatic changes without at least some adaptive evolutionary response to their changing environment". Given such profound changes in distribution, species composition, extinction, adaptive response and evolution during the Quaternary climatic changes, the idea of something "undisturbed" before the influence of man is untenable. We may talk of unaltered or altered/changed but "disturbed" is an untenable human construct.

It is thus very anthropocentric to argue about disturbance and recovery; "disturbance" (which in itself is an anthropocentric construct as it implies an original preferred state) and change have been prevalent throughout geological history. In this context, the arguments about nature benign (or Gaia) versus nature linear are somewhat unhelpful, as are the concepts of succession and maturity (Clements and Shelford, 1939; Worster, 1994, Chapter 11) when seen on a longer time scale. The geological perspective suggests that there has simply been continual change in species existence, composition and dominance in response to environmental conditions. Embedded in "nature benign" is the notion of a preferred state (it will go back to what it was – which is beneficial to or preferred by us). Embedded in the notion of maturity is a logical progression to a (comforting) stable condition. The notion of succession imposes a comforting logic onto what is a continually changing, mutually adjusting system responding to external changes and internal states. Nature benign, nature linear, succession and maturity are thus cultural constructs which have appeared to deal with such a complexity. What is more useful is to think about responses to change in terms of prior adaptation. As Orians (1975) believed: "*ecosystem behaviour in response to perturbations depends primarily on the adaptive characteristics of the organisms in the system.* Since these characteristics reflect past histories of experiences with perturbations and the continual evolution of associated species, we need to understand these adaptations better if we wish to improve our predictive powers about the effects of perturbations" (emphasis added). For example, if the frequency of fires increases, those organisms already adapted to fire (with, for example, below ground buds, protected seeds, the ability to sprout from damaged tissue) will flourish relative to those which do not have such adaptations. The system is thus responding internally to external changes but there is also an internal response to these internal changes in a way that is more akin to Per Bak's "*self-organised criticality*" (emphasis added) (Bak, 1997). Here, *nature is seen as perpetually out of balance but organized in a poised, or critical, state* when anything can happen but within well defined statistical laws. Minor changes can trigger events, which may be catastrophic, rather than events following a smooth gradual path. *The state is established solely because of the dynamical interactions among the individual elements of the system.*

If we accept that our view of nature is important because it will influence the way in which we treat and attempt to manage it, *how then should we view nature?* A bottom line is that the system is *mutually inter-reactive*. Change may occur in a system because of the mutual interactions, independent of external changes. External changes may also trigger shifts in state. The difficulty arises when one tries to see simple cause and effect relationships in that the relationships are not merely mechanistic between components because the components themselves can change.

This is a key point that Holling *et al.* (1996) make in their chapter on self-organization in ecosystems in Walker and Steffen (1996). Here, there is the point we have made earlier about interconnectedness with "hundreds to thousands of species interacting among themselves and their physical and chemical environment". Then they make an important assessment "But *not all those interactions have the same strength or the same direction*" (emphasis added). It is a limited number of species which are involved in the set of processes which structurally determine the functional diversity and not all species: species may be divided into "drivers" and "passengers" a small number of biotic (and abiotic) variables whose interactions form the "template" while others "go along for the ride" as it were, being affected by the ecosystem but do not, in turn, notably affect the ecosystem. As an example, they cite the relationships between tree species and budworms which feed on the trees – if unchecked the budworms will defoliate the trees, leading to forest decline; however, the budworms are eaten by insectivorous birds and it is a set of 35 such species of birds which eat the budworms which are key "drivers" in ensuring the continuity of the forest. Thus while many other species also exist in the forest it is the relationship between the trees, budworms and birds that is seen as a critical one; mosses, for example, may grow but they are not "drivers". They then make the second key point: "*The driver role of a species may only become apparent every now and then under particular conditions that trigger their key structuring function*" (emphasis added). Thus the critical components may change in the face of external change and only a small set of self-organizing processes are critical in forming the structure and function of the ecosystem (though this in turn is posited on the assumption of preferred states of ecosystems and has recovery embedded in it). Identification of this set would therefore appear to be a priority if we are interested in, or prefer, that state of an ecosystem that is maintained by that set.

Self-organization also appears to be a key concept in the view of Pahl-Wostl (1995), with the view that *living systems can create new types of responses to unprecedented situations and we should enable the capacity of nature systems for self-organization*. Certainly Drury (1998) in his thought-provoking book on *Chance and Change: Ecology for conservationists* challenges conventional ecological approaches, arguing that *nature works on the basis of one-to-one species interactions, variability and chance*. He makes the distinction between what "is" (what you can go out and see for yourself) and what "ought to be" (theorems of traditional ecology). I would argue that it is difficult to see what "is" in a way that is totally free of preconceptions, but it is interesting to note that the author writes that what he has seen "during decades of fieldwork is *neither pervasive order, nor chaos, but comfortable disorder*" (emphasis added) arguing that disorder is what makes the world work. The construct of disorder is a further step away from the concept of self-organization. However (even though underpinned by the assumption that species are important), his argument becomes more persuasive when he writes that "*individual organisms facing an uncertain future cannot afford consistency*" adding that parents producing more varied offspring have a better chance of being represented in the next generation than those who produce uniform batches of young – especially if environmental conditions change.

Certainly, in this book where we are discussing our attitudes to and concepts about nature in the face of environmental change we can see that the proposition that the relevance of either ecosystem science or notions of stability are severely contested. Kirkman (1997) goes so far as to say that "environmental thought might do well to divest itself of its ecological commitments". The author proposes that ecology cannot be all things to all people and that ecological concepts (from ecological science) can be appropriated (ecologism as a world view or ideology) to inform speculation, arguing that "its (scientific ecology) concepts cannot be readily made to serve purposes outside of their specialised contexts without a loss of meaning". Lackey (1998) notes that while "advocates glowingly describe ecosystem management as an approach which will protect the environment, maintain healthy ecosystems, preserve biological diversity and ensure sustainable development" the definitions of such management are vague and clarify little. The author indeed proposes "seven pillars of ecosystem management", only one of which involves scientific information:

1. Ecosystem management reflects a stage in the continuing evolution of social values and priorities; it is neither a beginning nor an end.
2. Ecosystem management is place-based and the boundaries of the place must be clearly and formally defined.
3. Ecosystem management should maintain ecosystems in the appropriate condition to achieve desired social benefits.
4. Ecosystem management should take advantage of the ability of ecosystems to respond to a variety of stressors, natural and man-made, but all ecosystems have limited ability to accommodate stressors and maintain a desired state.
5. Ecosystem management may or may not result in emphasis on biological diversity
6. The term sustainability, if used at all in ecosystem management, should be clearly defined – specifically the time frame of concern, the benefits and costs of concern, and the relative priority of the benefits and costs.
7. Scientific information is important for effective ecosystem management, but is only one element in a decision-making process that is fundamentally one of public and private choice.

The key elements are defined as "the application of ecological and social information, options, and constraints to achieve desired social benefits within a defined geographic area and over a specific period" and that appearing to "best respond to society's current and future needs". I guess that this is yet another way of saying "Ecology for People".

Finally, I would cite the writings of Williams and Patterson (1996). Here a person is seen as a "social agent who seeks out and creates meaning in the environment". This is paralleled by May (1996) who, although writing of tourism and authenticity, makes a plea that, rather than "forcing experience through the restrictive categories of ideal types" we should be "capable of mapping the ambiguities of individual experience". I come back to Steinbeck: "a man looks at reality and brings to it his own limitations". From all this I extract the sense that:

1. Concepts and constructs are important because the way we view nature influences the way we treat it.

2. Concepts and constructs are never "right" or "wrong" (all can be contested, and evidence found for them or against them), because they relate to individual meaning.
3. Education has inculcated ideas of ecosystems, stability, recovery and succession.
4. If these are now contested, what should our more "appropriate" constructs be? It is almost a case of: "Pick a construct – any construct", from the deck available. What do you chose? Order? Chaos? (What use is that construct?) How do you choose – with reference to what – ideology?
5. Currently "appropriate" constructs might appear to be related to: social benefit (both material and spiritual), the ability to cope with change, and the inextricability of nature and culture.

Change is the topic of the next chapter and subsequent chapters will consider wilderness and values of nature, soil resources and managed and domesticated ecosystems, before we can get nearer to working out some more appropriate constructs in the concluding chapter.

3.4 Questions

These first three chapters have been essays which consider some of the important factors which influence the debate about environmental change and human activity in the terrestrial biosphere. We have considered a range of ideas, including ecological concepts, the degree of human influence on ecosystems, values and ideas about how the world works. It is now appropriate to attempt to structure these themes in a way that might be useful when considering specific ecosystems. The headings I have grouped in our discussions under are as follows:

- Concepts
- States
- Preference
- Function
- Actions

The proposal is that these are not necessarily sequential stages but it is necessary to examine these questions during the consideration of the relationships between environmental change, ecosystems, our constructs of them, and the human response to change.

Concepts. How do we see and approach the environment and ecosystems?

Given that the way we view the environment influences the way we treat it, what are our constructs, values and attitudes – and indeed what should our constructs be and how should we judge their appropriateness and by what criteria? Is nature benign, linear, chaotic, robust or fragile or some or all of these things?

States. How do we evaluate a situation?

What are the patterns, distributions and extents of observable phenomena? In the monitoring (e.g. remote sensing) and measurment of these, what indices do we use?

Function. How does the system work?

How do we see and describe functional linkages and predict responses – through modelling? What will the consequences of our proposed actions be? Will they actually lead to the desired state?

Preferences. What do we want from a situation?

What state do we prefer? What is the utility and material gain; the benefit? What is the benefit to spiritual welfare? How do we approach differences of opinion and degrees of agreement?

Actions. How do we go about reaching a desired situation?

How do we get to the desired state, including ecosystem and societal state? What are the roles of institutions, implementation, infrastructure and enablement? Do we undertake ecosystem manipulation, conservation and/or preservation?

Further reading _____

***	Essential reading
**	Key reading
*	Important reading
	Recommended reading
---

*   Cherrett, J.M. (ed.) (1989b) *Ecological Concepts: The contribution of ecology to an understanding of the natural world.* British Ecological Society/Blackwell.
**   Golley, F.B. (1993) *A History of the Ecosystem Concept.* Yale University Press.
    Lackey, R.T. (1998) Seven pillars of ecosystem management. *Landscape and Urban Planning*, 40, 1–3, 21–30.
*** Pahl-Wostl, C. (1995) *The Dynamic Nature of Ecosystems: Chaos and order entwined.* Wiley.
*   Pimm, S.L. (1991) *The Balance of Nature? Ecological issues in the conservation of species and communities.* Chicago University Press.
    Sheail, J. (1995) The ecologist and environmental history – a British perspective. *Journal of Biogeography*, 22, 953–966.
**   Shrader-Frechette, K. and McCoy, E. (1993) *Method in Ecology: Strategies for conservation.* Cambridge University Press.
    Tansley, A.G. (1935) The use and abuse of vegetational concepts and terms. *Ecology*, 16, 284–307.

Williams, D.R. and Patterson, M.E. (1996) Environmental meaning and ecosystem management: perspectives from environmental psychology and human geography. *Society and Natural Resources*, 9, 5, 507–521.

** Worster, D. (1994) *Nature's Economy: A history of ecological ideas.* Cambridge University Press.

Zimmerer, K.S. (1996a) Ecology as cornerstone and chimera in human geography. Chapter 6 in Earle, C., Mathewson, K. and Kenzer, M.S. *Concepts in Human Geography.* Rowan and Littlefield, London.

Chapter 4

Ecosystems, society and environmental change

Summary of key points:

1. Ecosystems are valued according to social preferences.
2. Ecosystem components can change in the face of environmental change in terms of areal extent, structure, species composition, function and adaptations which may be physiological and behavioural and according to predisposition.
3. Social preferences about ecosystem states are expressed in terms of utilitarian and spiritual/intrinsic value where the criteria include putative "naturalness", capacity to support preferred species and products of commercial value.

4.1 Societal goals and ecosystems

One evening when the stream ran loudly
I went into a wood with two friends
Whose differences and arguments were genuine.
I had no idea of what they were hunting.
One shouted out that the trees were glowing;
The other disagreed, insisted
Nothing had begun yet.

Brian Patten (1971) *The Irrelevant Song*. George Allen & Unwin

We can break down the steps in relating social goals to ecosystems as follows:

1. Ecosystems will react to external changes in that components and flows will alter in a mutually interactive way.
2. Both gradual changes and abrupt transitions can occur.
3. Species can change in their distributions through prior adaptations. They can also evolve new adaptations, including physiological and behavioural adaptations.
4. We have a number of conceptual models of ecosystem behaviour. These models, as is evident from Worster's (1994) *Nature's Economy: A history of ecological ideas*, are essentially cultural constructs. Even though evidence can be found to support such conceptual models an element of surprise may occur if they are used to predict the outcome of management actions.
5. We might then seek a way forward by using those conceptual models that we have, but realize their limitations and try to specify the uncertainties.

6. Such information about limitations and uncertainties should be democratized, that is, made available to all those implicated in the effects of management actions.
7. Social significance is placed upon perceptions of the changes in the state of the environment and ecosystems and also on the current body of ecosystem knowledge.
8. Socially defined goal agreement is just as an important an issue in guiding ecosystem management as ecosystem models are.

In talking about achieving goals, the human needs of material welfare (security, shelter, food) and spiritual welfare (appreciation of the environment) are involved. Concepts such as productivity, adaptation, competition, disturbance and recovery, the succession of one vegetation type to another and of maturity are useful in guiding our thoughts about what might happen with and without human intervention. We can then judge present and predicted ecosystem states in terms of human needs, while stressing that these are not purely utilitarian, in terms of ecosystem products, but are also ethical and aesthetic. Conflicts can arise, of course, where utilitarian and spiritual needs each suggest different forms of management (exploitation versus conservation) but the way ahead is to seek common goals between utilitarian and spiritual needs. These are not necessarily mutually exclusive and indeed conservation policies which also have utilitarian benefits can be the most successful.

In these contexts, sustainability and biodiversity are the current buzz words. Sustainability is the topic of many books and much discussion, but at the very least it means being able to continue provision from ecosystems rather than decreasing the ability of ecosystems to provide material and spiritual goods. Biodiversity is also subject to much discussion (e.g. Mooney *et al.* (1996) in *Functional Roles of Biodiversity*). The current orthodoxy is that genetic diversity is inherently important in the way ecosystems can respond to change. If there exist only a few species, closely adapted to the environment which they are in and then the environment changes, then species extinction (assumed to be regrettable) is liable to occur. This is not necessarily defensible in that Huntley *et al.* (1997) show that migrations from other areas occur and indeed for some species, environmental change can lead to the stasis of some adaptable organisms. Biodiversity as a societal goal in itself is not necessarily easily defended – it is merely a numerical exercise akin to stamp collecting – the more different things you have, the better. This notion is certainly challenged by Perlman and Adelson (1997) who ask: "If biodiversity is good, is more biodiversity always better than less biodiversity?" together with "Does more different mean more important?" They conclude that there is no one right answer but provide possible answers in terms of context, for example, more may be seen as less desirable in terms of introduced species. They also argue that species counts in themselves are misleading and thus if there are, say, 10 species of animal type x (e.g. birds) and 1 of animal type y (e.g. reptiles) this might be preferable to having 12 of animal x and none of y, even though the latter situation has a higher total species count. In addition, a high numerical count of species is not necessarily defensible in terms of function as many species can perform the same tasks. In terms of ecosystem stability it cannot be defended as many ecosystems have already changed radically through geological history. Also, in terms of interdependency and components, high species counts might be more defensible as an ecosystem is no more than the sum of its components, but again we have seen the components can change and evolve. It is in the end, mainly a preference for diversity in the sense of "richness" and an ethical argument: despite all the extinctions of species in the past (Figure 1.1(b), p. 6) many people feel it is not right for people to be responsible for extinctions. In addition, the current rate of extinctions in relation to human activity appears to be much more rapid than during the geological past. This, and the fact that biodiversity conservation may have utilitarian benefits, is why biodiversity is something upon which many people agree.

In short, we actually only have one goal: self-interest. In the face of environmental change, our endeavour is therefore to preserve our self-interest in a material sense for our bodily well being and in an ethical sense for our spiritual well being.

4.2 Environmental changes – what is involved?

As I said in Chapter 1, a basic message is that it is certainly not the case that we have a fixed set of climate and vegetation states at the moment. Yet, we are trying to predict future changes from a fixed base line. Mitchell and Hulme (1999) wrote that "The inherent uncertainty in regional climate prediction is such that scientists must counter the damaging norm of assuming a stationary climate". They also stress the cascade of uncertainties (Figure 4.1) and state that conventional understanding has hitherto adopted a 30-year mean and so

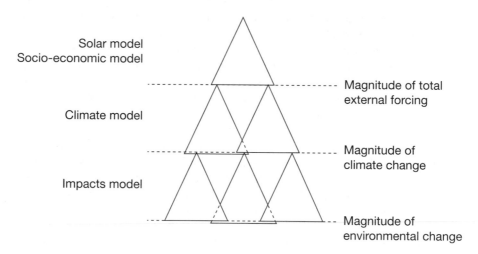

Figure 4.1 The cascade of uncertainties (Mitchell and Hulme, 1999). The range of uncertainties about external forcing magnifies as we predict the magnitude of climatic change and successively to the range of possible impacts.

humans are only adapted to the latest 30-year mean. Climates always show variability and trends, as Anderson and Willebrand (1996) show in their book on the dynamics of decadal climate variability and as Burt and Shahgedanova (1998) show for records at the Oxford Radcliffe Meteorological Station since 1815 (Figure 4.2). It is thus a case that during a scenario of continual past and present change, we are trying to understand the nature of these changes and think how we might predict future changes. Mitchell and Hulme feel that the important endeavours are to acquire a better understanding of climatic variability and the sensitivity of natural environments and human societies to that climatic variability.

There is, in fact, very little completely natural vegetation which exists today (Goudie, 1999). Westhoff (1983) sought to classify vegetation types in terms of the degree of human impact:

- Natural – not affected by human activity.
- Subnatural – essentially the same, but some influence.
- Semi-natural – spontaneous growth but some influence and alteration.
- Cultural – essentially influenced by man.

There thus is a category of "Natural" which is "untouched", though it is doubtful whether even the vegetation of high mountains and other remote areas is not touched to some degree, however slight and indirect, by human activity. This may be solely by anthropogenic particles, gases or other chemicals in the atmosphere rather than a direct influence.

It is useful, indeed, to classify the impacts and they can be seen as:

- Direct – e.g. by deforestation, planting; or
- Indirect – e.g. by air pollution.

Whether direct or indirect the impacts may involve:

- Replacement – e.g. by crops or non-vegetation land use; or
- Modification – e.g. by changing species composition or changing growth rates.

Simmons (1982) has classified the effects as:

1. Domestication – development for human use (e.g. crop cereals from grasses).
2. Simplification – fewer species than originally.
3. Obliteration – extinction of species or loss of vegetation type at a site.
4. Diversification – greater number of species than originally.
5. Conservation – protective management of existing species and vegetation types.

Sometimes there is a "downstream" effect which can be a non-spatial "knock-on" effect or a consequence of spatial proximity. Here because of, for example, water flow from land to streams or air flow from industrial areas to other areas, substances legitimately used in one activity become present in downstream or down wind areas where they are unwelcome, deleterious and regarded as pollutants. Another element is of dispersion of the effect – a much wider

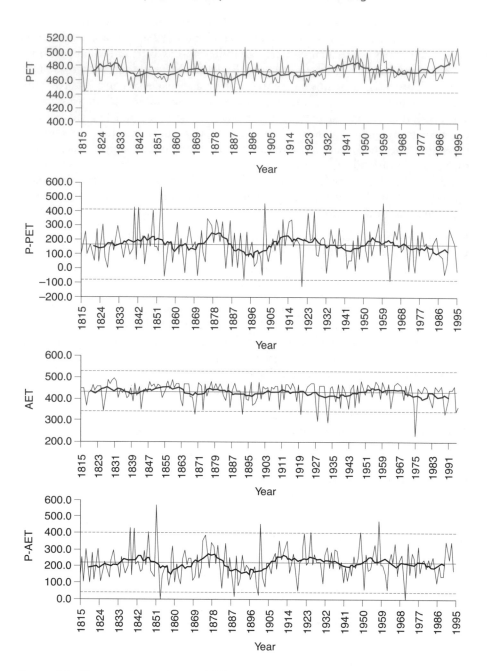

Figure 4.2 Climates always show variability and trends as Burt and Shahgedanova (1998) show for records at the Oxford Radcliffe Meteorological Station since 1815. PET = Potential Evapotranspiration, P = Precipitation, AET = Actual Evapotranspiration. All data are in mm. Heavy line is the running mean.

reaction to individual activity than might be predicted and often of a cumulative nature.

We thus need to understand the nature of these "downstream" and dispersive or cumulative effects. We also need to know, however, what the biosphere response is going to be. Some aspects of the biosphere may be able to react to, and absorb the nature of any effects without significant deterioration. The terrestrial biosphere may thus be unchanging in the face of change or highly sensitive and reactive to change. In terms of our resources of soil, plants and animals, they may be renewable, or the changes may threaten renewability over time.

It is important to examine the nature of changes, the responses of organisms to change and to see current changes in an historical perspective, rather than just assuming that "change" is a uniform phenomenon and that changes are necessarily detrimental, with consequences that need to be, and can be, tackled.

If we attempt to classify changes, the timescale of change is an important factor. In choosing a timescale, we might naturally select periods which have significance for humans, such as a generation, a year or a few days. Relative to the timescales important to people, changes which occur on geological timescales (thousands to millions of years) are almost imperceptible, though they may manifest themselves in terms of earthquakes indicating adjustments at tectonic plate boundaries or rock falls or landslides which result from the accumulated effects of gradual processes. We tend to perceive changes such as floods, storms and daily weather as far more significant because their effects are readily apparent on the hours' or days' timescale. Intermediate are perhaps those effects which are apparent in a life span such as coastal erosion and deposition or changes in vegetation over a number of years, where past memories can be used to contrast with the current situation.

The effects of the more gradual, more imperceptible changes often give rise to more debate simply because the evidence for change is often clouded by shorter-term events. However, these are often the ones which there may be more concern about as the less perceptible changes may in the long term be far more wide-reaching. Changes can thus be seen in terms of timescales relevant to people.

Changes may be:

1. Long term, gradual or almost imperceptible (hundreds, thousands or millions of years), e.g. climatic change, continental drift.
2. Medium term, life span changes (tens of years), e.g. increases in air pollution.
3. Short term (hours, days, weeks), e.g. dumping of waste.

Each of these may be:

(a) Continuous (especially (1) and also (2)).
(b) Intermittent – confined to a point, or points, in time (especially (3)); and/or
   (i) Temporary – returning to the former state subsequent to the change (e.g. an accidental, short lived spill of biodegradable waste);

(ii) Sustained. Here, the change may then itself be:
 – Stable (e.g. building of a factory which continues to produce pollutants new to the area but which subsequently do not change over time); or
 – Progressive (e.g. a gradual drying out of an area following ground-water abstraction);

(iii) Fluctuating (e.g. patterns of variations in yearly rainfall without a sys-tematic trend).

These are illustrated in Figure 4.3. Changed states may also pass thresholds where the effects become more, or less, evident. Examples of the timescales of changes are given in the book on decadal climatic variability by Anderson and Willebrand (1996). These include reconstructions of the paleo-record and for the more recent period of instrumentation.

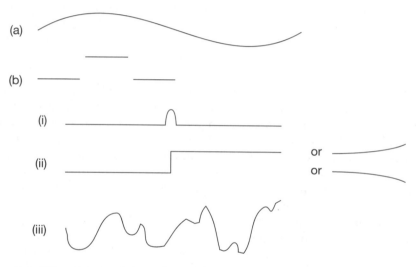

Figure 4.3 Different types of change (see text).

4.3 How does the biosphere react to change?

It is clear that the distributions of vegetation types can be plotted in relation to simple, time-averaged climatic variables (Figure 4.4). However, it is far from the case that we can simply imagine a shift of climate and a consequent shift in vegetation types (or agriculture). There are many complicating factors. We should begin by considering pre-existing conditions, sequences and histories.

If we start by looking at studies of biotic responses to environmental change in the past, Godwin's (1975) *A History of the British Flora* provides some inter-esting insights. The author concludes that there is "no evidence for the pro-gressive loss of species through successive glacial stages". Species recorded by means of pollen identification in interglacial deposits re-occur between each glaciation. Also during interglacial periods, species characteristic of full-glacial

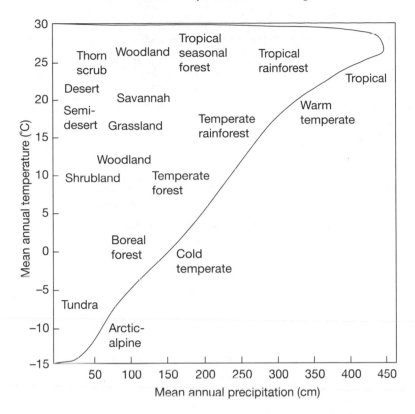

Figure 4.4 World vegetation types in relation to climate (from Houghton, 1997). The implication is that if climate changes, the spatial distribution of these types will also change, though the seasonal distribution of temperature and precipitation will be as important as the annual means.

periods were also present, either in mixed assemblages with variable conditions (including colder areas within warmer ones and/or in "refugia"). During glacial periods species distributions changed with distributions moving away from ice advance but returning with ice retreat. Note that, at this stage, these are "value free" statements, though the notion of "recovery" can be readily embedded and we may derive some comforting notion of return after glacial "disturbance" if we so wish. Godwin constructs a logical sequence of vegetation changes in an interglacial period (Figure 4.5).

If we can learn from this, two elements are important. Firstly the response to climate and, secondly, that the sequence itself is important – the vegetation and soils relate not only to the stages but also to the pre-existing state. Thus one trend is that plants associated with the climate which is current at any one stage tend to dominate (low herb and open shrub when cold, deciduous forest when warmer) but also there is a progression of soil nutrient decrease. Glacial deposits can be nutrient-rich, becoming less so with age through leaching. This leaching is especially marked as early cool conditions were often drier while the later

The Interglacial Cycle				
Characteristics of	Cryocratic	Protocratic	Mesocratic	Telocratic
Climate	Cold	Warm	Thermal maximum	Cooling
Soils	Immature, unstable, base-rich	Fixed but transitional	Brown earths	Podsols and blanket-bog
Vegetation	Open herb and low shrub	Park-tundra to light wood	Closed deciduous forest	Coniferous woodland and acidic heath
Floristic elements	Arctic and alpine	Residual arctic-alpine, steppe and S. European, weeds and ruderals	Woodland plants and thermophiles	Recession of thermophiles

Mean temperature (curve shown across top of table)

Figure 4.5 Vegetation changes in the last glaciation (Godwin, 1975). There are relationships with climate, but this does not necessarily imply necessity, reversibility or predictability. It is far from the case that we can simply imagine a shift the climate and a consequent shift in vegetation types. There are many complicating factors especially pre-existing conditions, sequences and histories.

cooling phase was often wetter, further encouraging leaching and acidity and also encouraging peat growth in bogs. Thus with cooling after warming the soils are more acidic than in the initial cold phase and the return to colder conditions is characterized by plants associated with acidic (and wetter) conditions whereas the initial cold phase was characterized by plants associated with base-rich (drier) conditions. Temperature alone is not the only factor, moisture is also important and, in addition, the soil type has evolved to a different state. Thus we should think not only of *responses to different combinations of factors* but also of *progression*. While we have argued that continual change is the norm, and that we are not possibly facing change from a static starting point, it is to say that each change commences from a specific starting point and similar changes may therefore bring different responses because the conditions prior to each change can be different.

If we are then trying to predict the future responses we should therefore first look at responses to different combinations of climatic factors such as temperature and moisture (and various associated indices based on frequency of occurrence of key combinations and thresholds) – and also atmospheric changes as elevated carbon dioxide concentrations. Second, we should also look at the implications of these combinations for different starting points, including current soil and vegetation types. The first endeavour might be able to produce a predictive calibration between environmental change and biotic response. Here the past might be the key to the future. The second consideration makes prediction far more difficult and here the past is about as useful as last year's train timetable (that was how it (supposedly!) worked last year but it may not tell us anything about this year at all). So on the one hand we might have a mechanistic view of how nature "works" in terms of responses to change but we also have to encompass the view of non-linearity, irreversibility, progression and calibration for a variety of specific initial conditions. This is not the same as chaos nor does it particularly render attempts at deriving predictive relationship futile but it does mean we should be cautious about using the predictive relationships we derive and, if we believe them, endeavour to move to a second stage of seeing how they might apply to particular combinations at particular sites (Trudgill and Richards, 1997).

In attempting to derive predictive relationships, at issue here are the manifestations of change and their relationships between our assumptions about cause and effect. We may see a tree becoming defoliated (Plate 4.1), but are the leaves dying because of direct acid rain impact or from insect attack? If insects are visible, has acid rain weakened the tree and made it more susceptible to insect attack? Does insect attack make a tree more susceptible to acid rain? Has acid rain impoverished the nutrients in the soil, weakening the tree and made it more susceptible to insect attack? Is the tree merely suffering from drought? Some of these questions might be more easily answered than others, but it is clear, first, that any proposed course of actions depends on which answer appears to be correct and, second, that cause and effect may not be easy to establish.

Harvey (1969) has suggested that the cause and effect model has been successful but that this should not be the only model open to us in analysis and explanation nor is the real world necessarily governed exclusively by the operation of cause and effect. This is especially the case in complex ecosystems where manifestations could be coincidental, they will probably at least have more than one cause and there may be synergistic effects. This latter point will especially be the case if we see ecosystems as wholes which are required to have genuine properties which are not reducible just to the properties of components (see Chapters 2 and 3) – it is therefore highly unlikely that ecosystem manifestations will be understandable in terms of cause and effect linkages certainly with individual components (X, Y, Z) and even additively with a number of components ($X + Y + Z$) but more likely in terms of the consequences of the interactions between components ($X + Y = a$, $Y + Z = b$, $X + Z = c$) but which in itself may then not resolve into a simple $a + b + c$ as these three may similarly interact

Plate 4.1 Problems of cause and effect: defoliated trees on Mount Mitchell, N. Carolina, USA. Are the leaves dying because of direct acid rain impact or from insect attack? If insects are visible, has acid rain weakened the tree and made it more susceptible to insect attack? Does insect attack make a tree more susceptible to acid rain? Has acid rain impoverished the nutrients in the soil, weakening the tree and made it more susceptible to insect attack? Is the tree merely suffering from drought?

giving three more combinations and so on, the possibilities being immense with a large number of components.

To establish cause and effect, Innes (1991) suggests that:

1. there should be a strong correlation between the measured effect and the suspected cause;
2. the observed association should have a plausible mechanism (i.e. a reasonable explanation which does not contradict other known mechanisms);
3. we should be able to duplicate the effect under controlled conditions;
4. the cause should precede the effect; and
5. the weight of evidence from 1–4 should permit a convergence on a general agreement.

As the scale of the ecosystem increases, this all becomes more difficult, especially item 3 where there can be no replicate of a large system, each one having a unique history. The whole is not just more than the sum of the parts, causes become effects and effects become causes in unique temporal conjunctures of sequences of events.

The main difficulty is that one is first being dissective (identifying components) and second trying to aggregate the components to see how changes are manifest and suggest possible causes. So rather than being dissective and focusing on cause and effect in complex situations, it might be more useful to think about the nature of possible changes. Consider a woodland (Plate 3.2, p. 66): trees do not move around in response to conditions changing but animals often can. Each tree can therefore only have a *local* effect on its immediate surrounding area (evaporation, interception of water, shading, dropping leaf litter) and its surrounding area can have an effect on the tree (water supply, nutrient supply, growth rate, physiological response) and these can be distinguished from more *diffuse* effects which might be additive local effects (woodlands affecting runoff together with water quality and, through evaporation, rainfall; climate – additive daily weather, and site history – additive through time, affecting general tree growth) and also more *footloose* effects which need not necessarily be located at any one particular site but may be located at any number of possible sites where suitable conditions exist (grazing of vegetation which can be ubiquitous, gathering of fruits, nuts, seeds and other animals activities). Ecosystem manifestations are therefore more liable to be manifest in terms of the aggregate, diffuse effects and in terms of the footloose effects which are not necessarily related to location and be expressed in terms of quantity (extent, numbers, rates over time) or behaviour. It is also true that quantities and behaviour may be as much related to internal synergies as to external conditions. Many changes may be most readily discernible through changes in land use cover (Turner, 1994).

In more detail, manifestations of change can be grouped into several categories:

1. Areal extent in the horizontal direction – cover of vegetation/numbers of animals.
2. Structure in the vertical dimension – largely in relation to the presence or absence of different levels such as trees, shrubs, ground cover.
3. Species composition – relative frequency and abundance of different plant and animal species at a site/in an area (often in relation to advantages or disadvantages of pre-existing adaptations or the increase in a species not particularly previously manifest but which benefits from the changes, as under item 5, below).
4. Function. This includes:
   (a) physiological rate changes, including rates of photosynthesis in plants and respiration in plants and animals;
   (b) functioning of food chains including vegetation–herbivore relationships and predator–prey relationships;
   (c) cycling of (i) energy, (ii) water and (iii) nutrients.

5. Adaptations which may be behavioural or physiological. Responses here include different behaviour patterns in animals (e.g. migration or other movement), new responses not previously exhibited or the increase in an activity previously not particularly manifest and includes lifecycle changes and breeding rate.

So far, these assessments can be "value free" but inevitably they become loaded with our judgements and preferences so that we have to add a further consideration of significance. We might here have to take care to distinguish between non-loaded and loaded terms as follows.

Relatively value free terms	*Loaded terms*
change, alteration	unstable, collapse, disturbance, succession
unchanging	stable, equilibrium, undisturbed, pristine, natural, untouched
decrease	deterioration, loss, impoverishment, paucity
increase	outbreak, overrun, out of control, plague, infestation, augmentation, productive
new, introduction	invasion, alien
diverse*	diverse, rich

* This might be a numerical statement or a loaded term of preference

A further manifestation of change might be categorized as follows.

6. Significance in terms of human preference and/or value in both spiritual and utilitarian terms. This tends to be judged in terms of preconceptions of naturalness and any interests as well as negative feelings of changes which lead to losses seen as regrettable. Included from the above can be judgements on: 1 – extent (which is often the most noticeable manifestation); 2 – structure especially in landscape terms (with reference to pre-existing conditions and preferences such as shade); 3 – species loss or situations constructed as involving "invasion" or "alien" species (especially loss of rarities or other preferred species and relationships with perceptions of biodiversity); 4 – functions in relation to preferred states (especially in terms of productivity); and 5 – changes in distribution or behaviour which are notable and probably constructed as indicators of change (unusual occurrences outside pre-existing ranges, especially with the more mobile "warmer", "southern" insect or bird species seen in northern areas, or unusual behaviour, such as earlier breeding, flowering or leaf opening).

The manifestations vary in their tangibility, with 1, 2 and often 3 and 5 being visible, though 3 and certainly 4 and possibly 5 may only be measurable rather than immediately apparent. Clearly, some responses can only be seen through careful study and detailed recording or investigation, especially with the less visible species. Here value judgements are more esoteric and may be limited to small numbers of specialists who perform these studies. However, through communication, they can become a matter of more widely held orthodoxies of views both in the scientific community and in society as a whole. The more readily assessed responses in terms of cover (with associated landscape impacts), species abundance and unusual behaviour are more immediately subject to public reaction. Judgements of the significance of changes then relate to changes in preferred states and in constructions of indications of change. Cause and effect links are often implicit in these judgements.

However, rather than just thinking simply in terms of cause and effect, we should perhaps also think of functional linkages and the fact that the sensitivity of a state of a system to fluctuations in external factors depends on the importance of these factors to maintaining that state (Trudgill, 1988, p. 151). Change may not necessarily have an effect on the terrestrial biosphere: just because a change occurs, this does not necessarily mean that it is relevant to plants and animals in terms of items 1–5 above or significant in our judgements. This does not involve a reaction, more an irrelevancy (the change is irrelevant and the system is not dependent on the changing factor because it is not related to it in any way). Whether or not an effect is seen depends upon the nature of the precise linkages between the change and the system under study. Systems thus may or may not have a sensitivity to a change. This may include questions of proximity (local effects or their diffuse aggregates) and/or functional linkage (involving physical transfers of matter in the hydrosphere or atmosphere or in food chains and influences on behaviour). If the system has a sensitivity to a change, then the change may be said to have an *impact* on the terrestrial biosphere. The question then becomes one of to what extent any manifestation of interest depends upon any changing variables.

A critical distinction is then between direct and indirect impacts; these are usually interpreted as on-site and off-site impacts. A direct, on-site impact may include, for example, frost tolerance and changed frost occurrence at a particular site while an indirect impact may include, for example, subtle changes in wind directions meaning that there is a change in the distribution of air pollution from an industrial source which affects a forest some kilometres away.

Responses to impacts may be reactive or passive. In the former the change has an effect of triggering a response – the system itself changes through reacting to the change. In the latter, the system changes purely through the effect of the change on that system. For example, some plants may actually grow faster when disturbed – say by light trampling – in response to the disturbance while others may simply die off. The reactive response may lead to a new, changed system (for example, a different growth form of the pre-existing species or a new mix of species) or the maintenance of the pre-existing one.

The system may also be seen as showing a resilience, that is a resistance to change (usually with reference to a preferred state, which is often constructed as "original") or as showing tolerance which might be active or passive, with a counteractive response or an immunity. Either the change can be coped with or the system is not dependent on the changing factors. For example, an increase in rainfall acidity will not acidify a lime-rich soil as the lime will dissolve to *offset* the acidity and the system will be immune to the increased acidity – the status of lime-richness does not depend upon the nature of the rainfall but upon the existence of dissolvable limestone in the soil (Hornung, 1985). Some lichens are highly sensitive to air pollution (and will exhibit a passive response by dying out) while others show a high degree of tolerance to sulphur compounds in the atmosphere and can thrive quite well in cities even when pollution increases.

The system may thus be stable in the face of change either through reaction or immunity or it may be unstable through a reactive change. The responses of items 4 and 5 above may act to maintain the manifestations 1–3 as they were before a change or lead to a changed state.

Change may or may not be valued, but if it is not, then the qualities of robustness or fragility and susceptibility are seen as important. This may involve a physical or physiological adaptation to changed or changeable conditions or be related to the degree of productivity. As examples of these, tough grass stems can withstand trampling better than soft, squashable leaves; plants adapted to exclude heavy metals from uptake by roots can withstand increasing soil pollution from lead from car exhausts and faster growing plants with underground growing points may be able to resist trampling better than slower growing ones.

Many current writers have sought to research the relationships between environmental variables and biotic response and thereby predict possible outcomes of climatic change (e.g. Bolin *et al.* (1986) in *The Greenhouse Effect, Climatic Change and Ecosystems*, Breymeyer *et al.* (1996) in *Global Change: Effects on coniferous forests and grasslands*, Walker and Steffen (1996) in *Global Change and Terrestrial Ecosystems* and Shugart (1998) in *Terrestrial Ecosystems in Changing Environments*), together with the foundation of many other studies of plants in relation to environmental variables (e.g. Grime (1979), Odum (1989), Whittaker (1975) on communities and ecosystems and Usher and Williamson (1974) on ecological stability). There also now exist a number of journals addressing these broad themes, such as *Global Environmental Change* and *Climatic Change*. In general the ecosystem themes focus on four aspects of changes:

1. Areal – changes in range and distribution (biome response).
2. Internal – changes in internal composition and species frequency.
3. Ecophysiological responses including productivity – both generally in relation to elevated carbon dioxide concentrations and locally, according to changed conditions.
4. Losses – potential loss of species unable to migrate to more favoured areas, either in terms of the rapidity of any change, immobility or lack of facility to migrate (such as "corridors").

A typical paper might be that of Box *et al.* (1989) on "predicted effects of climatic change on distribution of ecologically important native tree and shrub species in Florida" which talks of a threefold effect: mortality of existing species in existing sites, species migration and colonization by exotics. Envisaging a 1 or 2 °C annual temperature increase and seasonal variations on these scenarios, the authors model the relationships with current climate and predict the pattern of vegetation changes in relation to envisaged climatic zone changes. Unsurprisingly, they predict a northwards movement of subtropical species.

However, the key interest here lies in the possible impediments to such a shift and the conclusion that species might need extensive human assistance to realize the potential of a newly expanded potential range, which is an important topic we shall return to in Chapter 8 when considering species conservation. Some writers are, in fact, optimistic: Davis (1990), writing on climate change and the survival of forest species in Woodwell (1990) states that while habitat fragmentation will result in the probability that many forest trees will not be able to disperse rapidly enough (see Table 4.1) to track changing climate closely (i.e. to spread and re-grow in more suitable areas), many trees have enormous geographical ranges and wide ranges of tolerance so that they may not become extinct. Kirilenko and Solomon (1998) also raise the possibility of new biomes which may appear which may be transitory and also represent degraded versions of what currently exist and which they term "depauperate" biomes.

Feedbacks between plant characteristics and site characteristics might also be important. Aerts (1995) suggests that evergreens growing on nutrient poor sites exhibit low rates of nutrient loss and low rates of litter decomposition which reinforces the low nutrient status of the site. The attendant slow growth, in turn, can lead to a low responsiveness to environmental changes and an inability to adapt at a site where warming occurs, leading to changes of distribution in the face of change (Burrows, 1990).

Much attention has also recently been given to the effects of elevated carbon dioxide concentrations (see Luo and Mooney (1999), *Carbon Dioxide and Environmental Stress*). This involves both the projections of changed growth but also of changing carbon stores. Wang and Polglase (1995) predict that with increased carbon dioxide and temperature rise, the balance of these two effects will mean that the tundra and boreal forests will increasingly emit more carbon while the humid forest will continue to store carbon. In terms of growth alone, Lee and Jarvis (1995) propose the general conclusion that growth will be accelerated but that the magnitude of this depends upon "tissue type, nutrition and environmental conditions". The authors also emphasize the roles of inter-species competition and that models should take into account regional soils and climate interactions, which stresses the need for calibration for local conditions. Additionally, Whittaker and Tribe (1998) rightly point out that changes in

Table 4.1 Examples of rates of migration of vegetation (summarized from the paleoecological literature by Kirilenko and Solomon (1998)

Vegetation type	Migration rate, years per km	
	Mean	Rapid
Tropical evergreen	1	0.1
Tropical deciduous	1	0.1
Temperate deciduous	5	0.5
Cool-temperate evergreen	3.33	0.2
Boreal evergreen	3.33	0.2
Northern boreal evergreen	3.33	0.2

species distributions will be first (and most) evident at the margins of the distributions.

Models which allow for spatial variation and patch-scale responses are proposed by Walker (1994) in a study of landscape to regional-scale responses to global change. This author sees the important factors in modelling as involving feedbacks via exchanges of energy, water and momentum as well as changes in biogeochemistry. All these processes are affected by ecosystem composition and the concept of plant functional types is used to classify the responses. A simpler approach involves the study of the limits of distributions of plants adopted by Chapin *et al.* (1993). Here the northern limits of distribution of the major biomes are seen in terms of plant tolerance to minimum winter temperatures. This readily facilitates predictions of changed distributions for a number of temperature change scenarios but again further detail has to be built in to allow for local variation in soil resources of nutrients and water.

The importance of local conditions and plant traits are stressed by a number of authors, often in the context of competitive success (Teughels *et al.*, 1995). Burke and Grime (1996) concluded, in a study of grassland, that plant community change was encouraged by the existence of bare ground and especially where disturbance was coincident with eutrophication. Rejmanek (1999) proposes that changes in distribution can actually be inhibited or encouraged by the conditions which plants themselves can create – a "positive neighbourhood feedback". There are two main stands of trees they consider – eastern hemlock (*Tsuga canadensis*) and sugar maple (*Acer saccharum*). Hemlock seedlings are unable to spread into areas of other hardwoods, such as red oak (*Quercus rubra*) as this provides a deep, thick, coarse litter layer on the forest floor which the hemlock seedlings are unable to penetrate and any seedlings can be smothered by leaf litter. By contrast, sugar maple seedlings are able to penetrate and survive in this litter. Sugar maple may thus spread into areas of red oak whereas hemlock tends not to. Where there are areas of white pine (*Pinus strobus*), the hemlock seedlings are better able to survive in the thinner litter. Hemlock thus might spread into areas of white pine rather than red oak. Low nitrogen availability under hemlock in turn leads to poor survival of the sugar maple under hemlock. In this way red oak becomes progressively mixed with sugar maple and white pine progressively becomes mixed with hemlock. The author feels that this denies the existence of a Clementsian climax but that the progression of vegetation changes is related more to what already exists, together with feedback relationships upon potential colonizers, rather than a simple progression to a single climax. These kind of observations highlight the importance of local conditions and the existence of current conditions in the study of plant response to climatic change. Huntley *et al.* (1995) stress the importance of the *potential* range of species in relation to climate – whereas the *actual* distribution will depend on site history as much as climate.

Huntley *et al.* (1995) then proceed to model changes in potential ranges in terms of climatic change seen, for example, for *Quercus ilex* (Holm or evergreen oak), currently regarded as a Mediterranean species with changes according to model predictions of warming (Figure 4.6). In terms of a sensitivity to

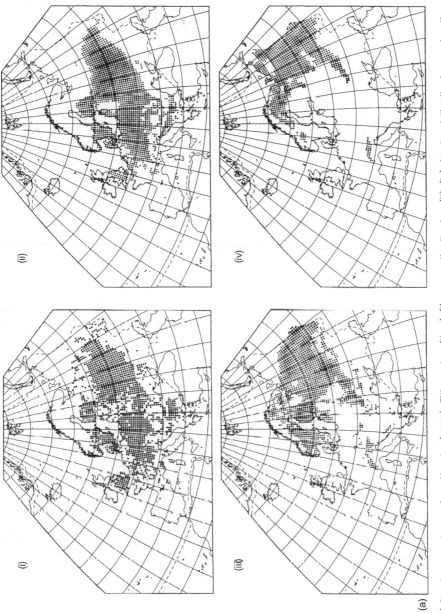

Figure 4.6 (a) Species spread maps from Huntley (1995). *Tilia cordata* (lime). (i) current distribution, (ii)–(iv) simulated distributions under climatic change.

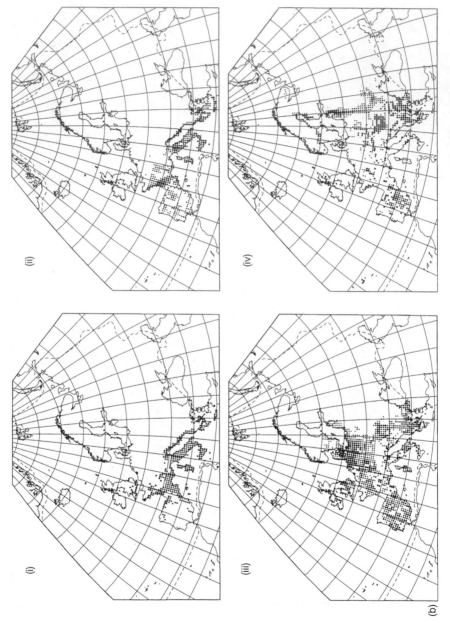

Figure 4.6 (b) Species spread maps from Huntley (1995). *Quercus ilex* (Holm oak). (i) current distribution, (ii)–(iv) simulated distributions under climatic change.

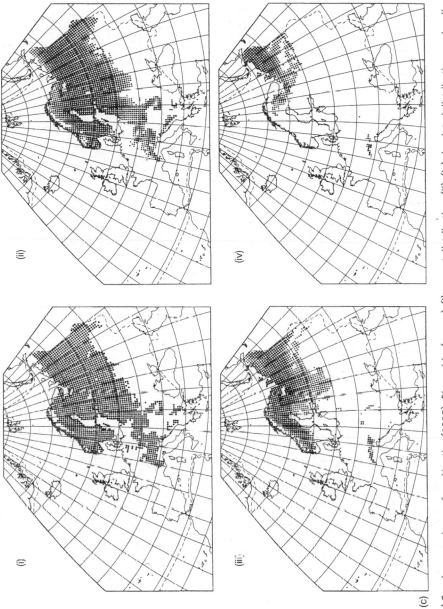

(i) (ii) (iii) (iv)

(c)

Figure 4.6 (c) Species spread maps from Huntley (1995). *Ficea abies* (spruce). (i) current distribution, (ii)–(iv) simulated distributions under climatic change.

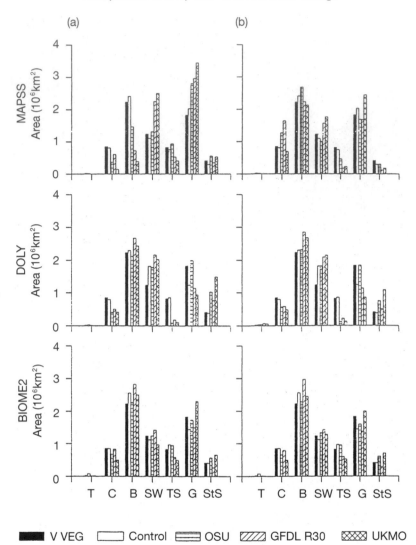

Figure 4.7 Comparison of area estimates under various climate scenarios (OSU: Oregon State University; GFDL R30: Geophysical Fluid Dynamic Laboratory; UKMO: UK Meteorological Office) as simulated by three different vegetation distribution models: (BIOME 2: Biogeography Model; DOLY: Dynamic Global Phytogeography Model) (a) without physiological CO_2 effects; and (b) with physiological CO_2 effects. The control simulations in Figure 2b do not include a CO_2 effect. The vegetation classes are aggregated from the original 21 V VEG types as follows: Tundra (T), 1; Conifer Forests (C), 2–4; Broadleaf Forests (B), 5–9; Savannah/Woodland (SW), 10, 11, 13–16; Subtropical Shrub/Steppe (StS), 12, 21; Temperate Shrub/Steppe (TS), 19, 21; and Grasslands (G), 17, 18.

change, Woodward *et al.* (1995) and VEMAP Members (1995) have been able to predict the physiological response to changing levels of atmospheric carbon dioxide, temperature and soil moisture though not the detailed response of vegetation composition (Figure 4.7). The key plant variables are photosynthesis, gas exchange through stomata, canopy conductance of water, canopy leaf area and the uptake of nitrogen. Stomatal conductance depends upon temperature and soil moisture; nitrogen uptake and maximum assimilation rates are proportional to irradiance and respiration depends in turn on nitrogen uptake and temperature. Increases in carbon dioxide can lead to a decrease in stomatal density in the context of adapting to limited water availability, improving water use efficiency (Beerling *et al.*, 1993) but on the other hand, increases in carbon dioxide generally stimulate photosynthesis and lead to more carbon storage in terrestrial ecosystems. However, there is one telling conclusion that "*various combinations of vegetation redistribution and altered biogeochemical cycles could produce scenarios ranging from increases in forest area and carbon sequestration to losses of forest area and losses of carbon stores*".

A further approach is that studies of the past will help to illumine the future. In a study of the Devonian era (see Table 1.1) Retallack (1996) writes of early forest soils and their role in past climatic change. He writes that forest ecosystems of that time have been viewed as engines of past climatic change. Here, he relates that the evolution of rhizospheres (roots) and increasing primary production (through increased photosynthetic use of carbon dioxide), together with increased silicate weathering all acted as sinks for carbon dioxide and all acted to reduce the greenhouse effect. The general weathering reaction is:

$$MSiO_4 \quad + \quad H_2O \quad + \quad CO_2 \quad \rightarrow \quad MHCO_3 \quad + \quad SiO_2$$

Silicate mineral   Water   Carbon dioxide   Carbonate   Soluble silica

where M = a metal cation such as Mg (magnesium)

It is thus appreciated that the weathering of silicate minerals uses up carbon dioxide producing soluble silicate and metal with carbonate ions in solution. The hypothesis is that the metal carbonates are released as nutrients for plant growth, thus facilitating further photosynthesis as well as using up carbon dioxide in the weathering process. The process is further aided by root respiration which itself produces carbon dioxide in the soil, aiding rock weathering. From evidence of palaeosols, which show increasing signs of weathering and biomass accumulation, Retallack argues that the increase in woody plants lead to greater carbon dioxide production leading to the subsequent ice ages at that time. If this reasoning is accepted, then it may well be predicted that the scenario of future increased lushness of plant growth posited by Robinson *et al.* (1998) (as discussed further, below) could be tenable, rather than an increased warming.

It is tempting to conclude that few things have exposed our ignorance more than attempts at predictive modelling, especially in view of the uncertainties about the changes as well as the uncertainties of the responses. However, Melillo *et al.* (1996), simplifies the approach in regarding the main constraints as related to the functional dependencies of plants upon temperature and water. The panel working on terrestrial biotic responses writes that cold tolerance ranges from the near unlimited cold tolerance of boreal deciduous trees in their leafless winter state to tropical trees, some of which are damaged by temperatures of less than 10 °C. At the polar treeline growing season warmth is seen as the limiting factor and so warmer summers and/or longer growing seasons would thus result in a poleward spread of cold-adapted trees. Warmer winters would also allow a poleward expansion of non-cold adapted trees currently present at lower latitudes than the cold-adapted species. Cold tolerant woody plants have evolved mechanisms to delay spring budburst until day length is adequate and/or a long enough chilling period has occurred; warmer winters might then force a retreat polewards of these species. Thus in the northern hemisphere, a general warming simply means the cold-adapted tree zone shifting northwards, while its southern boundary also shifts northwards as non-cold adapted trees expanding northwards into the zone previously occupied by cold-adapted trees.

In terms of water availability, rainforest trees can tolerate only a few weeks without rain (unless they can tap into deep groundwater sources) while other plants can store water, reduce losses or reduce their activity in dry periods. Clearly, it follows that changes in moisture availability will result in concomitant changes in plant distributions.

In this approach, there is little in the way of sophisticated modelling, more a statement of which plants grow in which climates and a simple prediction that if climatic zones shift, then vegetation zones will also shift. However, the details of the vegetation associations are not actually that easy to predict, especially in terms of structure and species abundance and of the history of changes relative to the previous states. Studies of vegetation changes in the past reveals that each successive interglacial, while all were characterized by increasing temperatures and better conditions for plant growth, exhibited different vegetation assemblages despite similar conditions (Watts, 1988). We might therefore fairly readily predict changes in vegetation types, in relation to their environmental adaptations, but less readily the species composition. In addition, as we said at the outset, information exists on vegetation changes, and thereby on the rate of spread of species as climate changes, through the study of the pollen records in dated cores of sediments. However, we have already pointed out that progressive changes are important as well as initial conditions. Additionally, as Melillo et al. (1996) point out, a knowledge of past dispersal rates might well not be a good guide to future rates for the following reasons. First, the modern, human-altered landscape may provide fewer dispersal sites and routes, resulting in slower dispersal while, second, dispersal rates may be greater in relation to the ways in which human activity acts to spread species both deliberately and accidentally. It is thus important to start from where we are, with often a fragmented vegetation cover, rather than where we might be. It is thus important to discuss these matters in terms of the different types of ecosystems – wilderness, domesticated and replaced. A knowledge of past responses to change will be more relevant in the former and progressively less so in the latter two.

It is highly likely that in the larger wilderness areas, the spatial arrangement of zones would change with some facility but in other areas where there are isolated areas of vegetation surrounded by landscapes dominated by continued human activity, species composition would change with less ease. Thus unlike say, woodland spreading over tundra in a contiguous area, with recruitment from the adjacent existing woodland, a woodland area surrounded by agricultural land would have few new recruitment opportunities. While the woodland may find itself, as it were, in a new climatic zone, the actual new vegetation would depend on the nature of the limited recruitment opportunities rather than automatically changing to the climatically-adapted vegetation zone. The biggest changes may well be seen in the cases where climatic change, like increased droughtiness, acts to decrease the impact of human activity, resulting from a decreased agricultural potential. Here, reversion from agriculture could see a spread of vegetation from contiguous areas.

In general, the important comparison is one of rates of climatic change and rates of evolution of animals and plants. Rhodes (1962) suggests that the average time of speciation in mammals was extremely slow and probably of the order of 500 000 years. Average change in diameter of early equid (horse) molars was less than 0.2 mm per million years while differences of 3 mm or more were present within single populations. For climatic changes on time scales from 10 to 10^4 years, it appears that for perennial plants, the evolutionary response producing new species is also too slow for new species to evolve in response to new climates. Instead changes occur in the local abundance of species (Type A response, Webb (1986) or in the geographic distribution of plants (type B response). In the latter, *dispersal leading to a change in spatial range has in the past appeared to be a relatively rapid process and, indeed, the main way of responding to change.* On a geological timescale the response appears to be almost instantaneous. The response of soil type, however, can show a considerable lag so that on the timescale of decades to centuries vegetation types may be found on soil types which they are not currently found (Kirschbaum, 1996). The most responsive fraction of the soil is the organic matter. Decomposition is limited by temperature and moisture effects upon the soil decomposing organisms (Kononova, 1966) (see Figure 4.8). Clearly, processes like respiration (and associated carbon release), denitrification and nitrification will increase if a soil with a high organic matter store is warmed with adequate moisture present. The other combinations (warmer, drier, colder wetter, colder drier) will tend to decrease biological activity (Figure 4.9). Again, the changes can be predicted to vary from wilderness areas to areas already changed by human activity.

4.4 Are the changes seen as detrimental? _____

Environmental change is the only constant factor in the history of the earth. The species which now exist in the terrestrial biosphere are those which have adapted to, or not been disadvantaged by, such changes. It remains, therefore, to examine the reasons why there is so much concern about perceived current changes. There are two perceptions of the significance of the current changes. One is the anthropocentric concept that they will be increasingly detrimental to

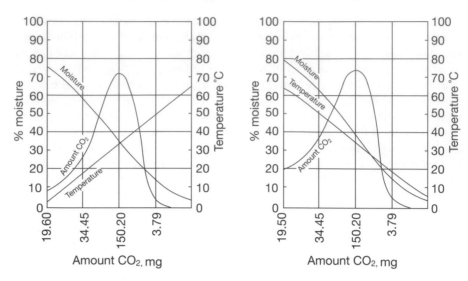

Figure 4.8 Ecosystem response and climatic conditions – combinations of moisture and temperature affect organic matter decomposition as measured by CO_2 production (from Kononova, 1966).

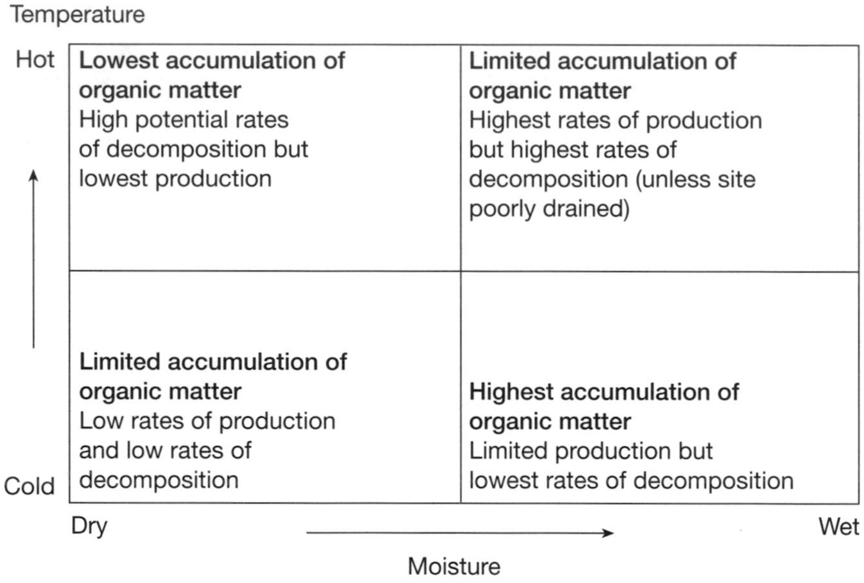

Temperature

	Dry	Wet
Hot	**Lowest accumulation of organic matter** High potential rates of decomposition but lowest production	**Limited accumulation of organic matter** Highest rates of production but highest rates of decomposition (unless site poorly drained)
Cold	**Limited accumulation of organic matter** Low rates of production and low rates of decomposition	**Highest accumulation of organic matter** Limited production but lowest rates of decomposition

Moisture

Figure 4.9 Organic matter decomposition and combinations of temperature and moisture.

human existence and the other is the less direct, more altruistic, concept that the changes will be increasingly detrimental to the life forms and environments on earth for which we have a feeling of stewardship. Equally, however, there are those who will deny that these changes are detrimental or they feel that they are not significant, either in the changes themselves or relative to other, more pressing concerns. Values may thus vary and attitudes may therefore range from concern to indifference.

The motivation for attempting to do something clearly stems from our definitions of detrimental. "Detrimental" is a relative description – relative to human perception, both in terms of the benefits to people and of a desired state which may be irrespective of any benefits. We may thus perceive something as detrimental if benefits are decreasing or not accruing in a way we expected. We may also see something as detrimental if a state does not conform to some desired, ideal state or concept even if benefits to us are not involved. It is thus perceived that changes will affect our well-being and the well-being of things we care about. There is also the point that if changes are perceived to be in some way detrimental, then the identification of the causes is a critical step. Broadly speaking if the changes are anthropogenic, then we may well be able to reverse any detrimental effect, but if they are caused by agencies outside our control, then they are far more difficult to influence.

The value judgements on changes as detrimental may thus be of:

1. Aesthetic/spiritual/moral preference – here we may regret the loss or gain of a species or a loss of some putative "naturalness" or preferred state (as discussed in the next chapter).
2. Decrease in the capacity of the environment to support a preferred state, species or product (e.g. in soil fertility or moisture status, as discussed in Part II).
3. Loss of commercially valued species or gain of commercially undesirable species.
4. Lowered productivity of commercially valued ecosystem product.

(Items 3 and 4 are included in the discussion in Part III.)

Because people tend to value ecosystems in their own vicinity or at a particular site, these judgements are more likely to be made on a locality basis, especially in relation to land tenure and management plans for sites, rather necessarily than on a regional, national or continental scale. In general, also, the changes may not always be judged as detrimental but simply as unwelcome since change in itself is not necessarily welcomed.

Some economic assessments of the costs of climatic change have been attempted (e.g. Demeritt and Rothman, 1999) but these can be highly contested, see the comment by Fankhauser and Tol (1999): "Demeritt and Rothman are simply wrong". While the bases and assumptions of speculative calculations can always be contested, the first two authors wished to make the point that the technical/economic assessments appear "scientific" (see Table 4.2) but should not be used to side-step the important moral and political debate.

In terms of specific standpoints, there seems to be a general sense that climatic change will be detrimental, however there are counter voices. Robinson *et al.* (1998) in a study of the environmental effects of increased atmospheric carbon dioxide feel that "Predictions of harmful climatic effects due to future

Table 4.2 Estimated costs of climatic change impacts (modified from Demeritt and Rothman (1999) after Fankhauser (1995))

		Area		
	$ loss per	EU	USA	China
Dryland loss	km²	2 000	2 000	0*
Coastal wetland loss	km²	5 000	5 000	500
Species and ecosystem loss	person	30	30	2
Forest value	km²	2 000	2 000	200
Water prices	m³	0.92	0.42	0.05
Value of statistical lives	person	1 500 000	1 500 000	100 000
Air pollution	kg	7.5	7.5	0.75
Migration	immigrant	4 500	4 500	1 000

* assumed to be defended

increases in minor greenhouse gases like CO_2 are in error and do not conform to current experimental knowledge" and that "the effect on the environment is likely to be benign. Greenhouse gases (here talking of CO_2) cause plant life, and animal life that depends upon it, to thrive. What mankind is doing is liberating carbon from beneath the Earth's surface and putting it into the atmosphere, where it is available for conversion into living organisms". The authors present "hypothesis 2", that there will be insignificant rises in temperature showing that while carbon dioxide has risen, temperatures have not increased markedly (Figure 4.10). They then present graphs of increased plant growth with increased carbon dioxide, pointing out the agricultural benefits of increased plant growth, helping to maintain and improve the health, longevity, prosperity and productivity of all people. "We are living in an increasingly lush environment of plants and animals as a result of CO_2 increase. Our children will enjoy an earth with far more plant and animal life than that with which we are now blessed."

The envisaged changes can thus be constructed in many different ways and the projections are no more than extrapolations of what we think we know. The

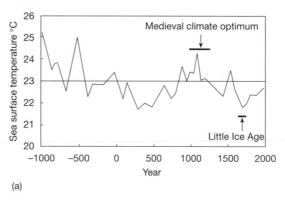

(a)

Figure 4.10 Refutations of global warming from Robinson *et al.* (1998).

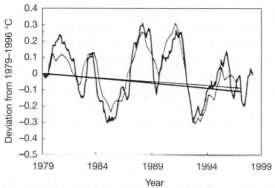

(b) radiosonde balloon (light line) and satellite (dark line) with trend lines

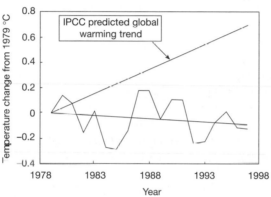

(c) tropospheric temperatures

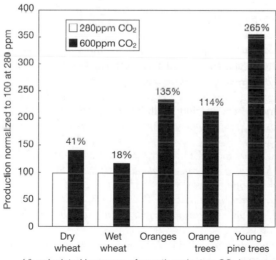

(d) calculated increases of growth under two CO_2 increases

Figure 4.10 (*continued*)

science of prediction has improved greatly. We have applied a lot of existing knowledge, and also learnt new knowledge, in the process of the investigations into the relationships between environmental change and the distributions of plants and animals, both in terms of functional physiology and behaviour. However, there are enough caveats about both the predictions of climatic change and of the biotic response to almost conclude that "we just don't know" what will happen. As we said in Chapter 1, this justifies a more fundamental look at capacity building, the ability to react to surprise, concepts of resilience and above all, at our value systems so we might be in a position to retain what we value and adapt our concepts and practices to changed circumstances. Thus, in Chapter 5 we discuss man–nature relationships in the contexts where the value is not dominantly utilitarian (as discussed in Part III) and look into the constructs and the values of wilderness, nature, nature conservation, urban living and gardening.

Further reading

> ** Key reading
> * Important reading
> Recommended reading

Berkes, F. and Folke, C. (eds) (1998). *Linking Social and Ecological Systems: Management practices and social mechanisms for building resilience.* Cambridge University Press.

Breymeyer, A.I., Hall, D.O., Melillo, I.M. and Argren, G.I. (eds) (1996). *Global Change: Effects on coniferous forests and grasslands.* SCOPE 56. Wiley.

Fairhead, J. and M. Leach (1996). *Misreading the African landscape: Society and ecology in a forest–savanna mosaic.* Cambridge University Press.

Houghton, J. (1996a) *Global Warming: The complete briefing.* Cambridge University Press.

* Jasonoff, S. and Wynne, B. (1998) Science and decision making. Chapter 1 in Rayner, S. and Malone, E. (eds) (1998) *Human Choice and Climatic Change: Volume 1 The Societal Framework.* Battelle Press, 1–87.

** Melillo, J.M. *et al.* (5 authors, 29 contributors) (1996). Terrestrial biotic responses to environmental change and feedbacks to climate. Chapter 9 in Houghton, J.F. *et al.* (6 eds) *Climate Change 1995. The science of climate change.* Cambridge University Press, IPCC.

Mooney, H.A., Cushman, J.H., Medina, E., Sala, O.E. and Schulze, E.F. (1996) *Functional Roles of Biodiversity. A global perspective.* SCOPE 55, Wiley.

* Rayner, S. and Malone, E. (eds) (1998) *Human Choice and Climatic Change: Volume 1 The Societal Framework.* Battelle Press, 1–87.

* Shugart, H.H. (1998) *Terrestrial Ecosystems in Changing Environments.* Cambridge University Press.

Walker, B. and Steffen, W. (1996) *Global Change and Terrestrial Ecosystems.* Cambridge University Press.

Woodwell, G.M. (ed.) (1990) *The Earth in Transition.* Patterns and processes of biotic impoverishment.

Chapter 5

Wilderness, nature and gardens

Summary of key points:

1. Wilderness is a concept as much as a place.
2. As a concept of a place with minimal human influence, there is a duality of being both valued and feared.
3. Nature reserve management is much about expression of human preference.
4. As such, much effort has hitherto gone into arresting succession.
5. In the future facilitating change will be a key issue to face, with habitat linkage an important factor.
6. Green spaces in cities and gardens are the "acceptable" face of nature, involving elements of (therapeutic) involvement but with control of nature.
7. In the face of environmental change it will be the range of culturally defined preferences for the range of nature experiences – from wilderness, reserves, parks and gardens – which will be critical in determining the outcome.

Nature is
an analogy
   a comparison of feelings
      a wholeness.
Nature is a hard thing
     callous and indifferent,
     a contrast.
Nature is a comfort,
     a human source of warmth,
     a sympathy.
Nature is finding yourself
Nature is being able to think
Nature is a vigorous force
     and participation
     joy, sweeping emotion
Nature is hope
Nature is individual
Nature is the inspiration.

Peace of mind is an Oak Tree in a Forest
Calm and unmoved
Strength is the solid mountain
Beauty is the wind's breath

Cool is the summer delicate bird weep forest.
Love is the touch of a sun's ray in the gloom
Calm is the blue bay sea by the marsh
and sand and shingle.
Stirring is the wind sound
Everything is the wind sound
so deep, inspire, instil
surge up
to fulfil the Mountain
quiet strength of character
giving peace of mind
calm, quiet, yet strong and upwelling
wistful life.
What do we want more than Nature?

Anywhere
any place
it moves, it stirs
be it sea or sun or stars or wind
it is Nature to feel the same
and to feel together
and balanced
and Nature within me and without

and beside me and fused and forward go
and sit in quiet content by the side of the sea.

Sometimes when I'm in the city
a cool fresh breeze
blows on my face and hair
and reminds me of where I should be
on the hills with the trees
out.

To re-find myself
like the shell being discovered
by the hermit crab.

There is much that man does love
that does not much love man
– the vast blue inhumanity
of boundless sea and sky

    I took a walk by the sea today
    I enjoyed it because
    the sea and the wind are bigger than
    me.

Vast inhumanity, of elements, of sky, of sea
makes my light,
my spark of life,
shine brighter.

> ... to come down off this feather-bed of civilisation and to find the globe granite under-foot and strewn with cutting flints ... To hold a pack upon a pack-saddle against a gale out of the north is no high industry, but it is one that serves to occupy and compose the mind. And when the present is so exacting, who can annoy himself about the future?
> Robert Louis Stevenson (1879) *Travels with a Donkey in the Cevennes*, Kegan Paul

5.1 Wilderness ecosystems

Wilderness ecosystems are those where the imprint of human activity is minimal and which are valued for their resources and their "naturalness". As such places like ice caps, deserts, mountain ranges and forests can be seen as wildernesses. They are essentially unmanaged and they can be with or without protective designation, such as a Protected Area classification. National Parks vary in their degree of management in different countries, and thus might appear as "wilderness" in this chapter, especially in the example of the USA (as discussed by Henderson (1992) who contrasts the North American and UK experiences) while indeed UK National Parks are perhaps so intensively used that they are really part of the domesticated landscape. Nature reserves and plantation forests are here regarded as part of the managed landscape. Wilderness then covers a category of non-domesticated land where natural landscapes, processes and products are given intrinsic value. Shafer and Hammitt (1995) cite the USA 1964 Wilderness Act as providing five descriptors:

- natural,
- solitude,
- primitive,
- unconfined and
- remote.

Utility is focused on the experience of "naturalness" or on the extraction of products with limited or minimal impact on the ecosystem. The cultural imprint is largely in terms of recreation, including footpaths and other access areas. Development such as access, recreational facilities and hotels is often the subject of some contention, as may be the practices which involve extraction.

One third of the global land surface is estimated to be in the wilderness category (McCloskey and Spalding, 1989), with 48 069 951 square kilometres in 1039 blocks of over 4 000 000 hectares (smaller blocks were not included); 41% of this amount is in the Arctic or Antarctic and 20% in temperate regions (Table 5.1, Figure 5.1).

5.2 Wilderness and nature values

There seems to be a move to value nature in monetary terms. Odum (1978) made a study of the value of wetlands in terms of the equivalence of the organisms in sewage processing; similar moves have valued nature in estate agents terms: "how much more would you pay for this view" and so on. Costanza et al. (1997) finally came up with an economic valuation of the world's ecosystems. This endeavour may well impress economists and is part of the endeavour to

Table 5.1 Wilderness habitats (after McCloskey and Spalding, 1989)

Vegetation type	km^2	%	No. of blocks
Tundra	20 047 533	41.7	100
Warm desert/semi-desert	9 329 531	19.4	389
Temperate needle-leaf forests	8 799 312	18.3	120
Tropical humid forests	3 006 855	6.3	77
Mixed mountain systems	1 973 391	4.1	76
Cold-winter deserts	1 478 494	3.1	51
Tropical dry forests	1 424 099	3.0	120
Tropical grasslands/savannahs	735 331	1.5	33
Temperate rainforests	450 215	0.9	15
Temperate broadleafed forests	290 646	0.6	20
Temperate grasslands	272 016	0.6	24
Evergreen sclerophyllous forests	170 885	0.4	7
Mixed island systems	91 647	0.2	7
Total	48 069 951		1039

Continent	km^2	% of total	% of continent	No. of blocks
Antarctica	13 208 983	27.5	100	1
N. America	9 077 418	18.9	37.5	85
Africa	8 232 382	17.1	27.5	434
Soviet Union	7 520 219	15.6	33.6	182
Asia	3 775 858	7.9	13.6	144
S. America	3 745 971	7.8	20.8	90
Australasia	2 370 567	4.9	27.9	91
Europe	138 553	0.3	2.8	11
Total	48 069 951			
World habitat	162 052 691			
% wilderness	29.7			

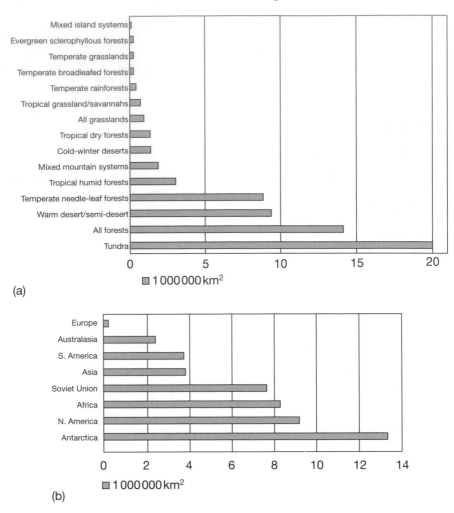

(a)

(b)

Figure 5.1 Wilderness habitats as defined by McCloskey and Spalding (1989). (a) Classified by vegetation types. Note the grasslands and forests are presented both as aggregate figures and by subdivisions of type. (b) Classified by continent.

explain the worth of nature to people who do not actually see it. However, I find this a rather bleak analysis and do not think we should be afraid to talk of an unquantifiable spiritual value and work towards the legitimacy of a plurality of values (Chapter 9, Conclusion) rather than forcing everything into monetary values. This latter can lead to a loss of meaning as described by Ashby (1978): "To quantify fragile values not only drains them of some of their meaning, it invests them also with a false meaning … and illusion of precision." Ashby places more trust in notions of altruism, sentiment and empathy. In a discussion of "Changing meanings of 'environment' ", Healey and Shaw (1994) also regret that the notion of *environmental care* has become replaced by *marketable assets* and have the aspiration that the sustainable development rhetoric may redress this. Booth (1997) simply says that: "if the benefits of preservation exceed the cost, then preservation should be undertaken, if not (ecosystems) should be exploited. However, if *ecosystems are valuable in their own right, and for this reason have moral standing, then the cost-benefit approach may no longer be appropriate*" (emphasis added). O'Riordan (1997) asks for a valuation as revelation and reconciliation – creating trusting and legitimizing procedures, rather than seeing things in terms of a willingness to pay. In these contexts, we should thus ask what wilderness and nature mean to mankind and assess how far the range of views might be shared amongst different groups of people and then lead on to discuss how we might be able to facilitate a range of views.

We all live in worlds of meaning and significance and thus there will always be many meanings and significances placed on nature (see Braun and Castree (1998), Escobar (1996), Evernden (1992), Darier (1999a), Rolston (1997) and also Merchant (1980) who lists nature as female, utopia, organism, disorder, mechanical order, mechanism, dominion and management; together with "nature writings" discussed by Bramwell (1989)). Evernden makes the useful distinction between different peoples' approaches to nature, one is "what is this *to* me?", with nature as extended self, and the other is "what is in this *for* me?" with nature as object. Some people will never see what others mean, so taking ecological evangelicalism to the extremes of placing a monetary value on nature may just help to convert some people but, in that endeavour, there can be a concomitant loss of meaning for others. Might we not celebrate the variety of views and concepts that pertain to different societies and cultures and also explore the range of meanings and then see which ones might be compatible in what places? We may have to decide on *dominant* meanings for different areas (wilderness reserve; productive agriculture, see also Chapter 9) but the "win-win" is where we can have a plurality of spiritually valued meanings about nature and the benefit of economically valued ecosystem products, whether by spatial separated and zoned sites or simultaneously at one site (see Chapter 1 with Barbier's discussion of multiple goals – biological, economic and social). So, we can ask here, what value systems pertain to wilderness, nature, nature conservation and to nature and gardens in urban areas?

Wilderness is as much a concept as a place. As a concept, it has changed. In the early history of people – and still for many people today – it was something to be avoided or conquered: terrifying, anarchic, lonely and hostile – the very antithesis and negation of civilization. A more recent view is of something valued for its intrinsic qualities, including the fact that it *is* relatively untouched by human beings. Here, the wilderness experience may bring the feeling of solitude and "being with the elements", giving spiritual uplift. This may be achieved through views from roads and car parks or, for the more physically active individual or groups, immersion and direct involvement. It may also be achieved more vicariously through television and films. Indeed, through television many people would now profess to cherish and want to protect wilderness without having ever been near any: "individuals in modern society love nature only in the abstract" (Banuri and Marglin, 1993, p. 17). Maybe also this might illumine why we are passionate about the rainforest even though most of us have not been there.

The trend from hostility to value has increased as the amount of wilderness has physically decreased, as we have become more technologically advanced and more able to cope with hazards and challenges and as we have become more urbanized and lost frequent contact with the environment. Nash (1973) writes in *Wilderness and the American Mind* how the first settlers in the USA regarded the American terrain as hostile, something to be tamed and conquered, but with abundant resources which could be put to human use (see *From Coastal Wilderness to Fruited Plain* – Whitney (1994)). Nash (1973) saw the shift in attitude in terms of extent – the original wilderness being seen by European settlers was so vast as to represent both something to be feared and something of an inexhaustible resource (citing the example of burning down a log cabin in order to extract the scarce metal nails). Felling then brought timber scarcity and that scarcity brought the idea of something to be valued. Conservation societies only arose at a later time when the society was substantially urbanized and the wilderness much reduced in extent and increasingly valued (Foster, 1997).

In a key writing, Cronon (1995b) in "The trouble with Wilderness or getting back to the wrong nature" called at that time for a re-thinking of wilderness. It should not be seen as a pristine nature – not as a place which stands apart from humanity, but it should be seen as a place of human creation – "a reflection of our own unexamined longings and desires". This much is evident from the title of Evernden's (1992) *Social Construction of Nature* who also dwells on some of the ambiguities involved (but see also the critique of Evernden in Quigley (1999) who writes of a "typical romantic longing" embedded in Evernden's discourse). In the context of wilderness as a construct Porteous (1996), in his book *Environmental Aesthetics*, shows an historical origin of the roots of our duality in our attitudes to wilderness as a place of both good and evil – and probably in either case as something awesome. Porteous writes of the Puritans' fear of the American wilderness "as a reflection for their disdain of sexuality. Wilderness, like sex was difficult to control, frenzied and not easily amenable to rules and regulations. The Puritans were looking for a second Eden, but were confronted, just beyond the pale of settlement, by a wilderness which was not only an actual threat to survival but also a dark sinister symbol of evil". Porteous contrasts this with an 1803 quote from "a gentleman explorer in Ohio", quoting from Nash (1973): "There is something that impresses the mind with awe in the shade and silence of these vast forests. In deep solitude, alone with nature, we converse with God". Interestingly he concludes that in the USA "the notion of wilderness as a positive aesthetic flowed westward from the eastern cities, never quite catching up with the trans-Appalachian pioneers who still stoutly maintained that the wilderness was howling, dismal or terrible and should be re-made so as to blossom like the rose".

It is, however, not an automatic tenet that while wilderness is unappreciated by those who have to live in or otherwise deal with it and only becomes valued by the city mind. The London writer, Dr Johnson, on travelling through Scotland looked with horror at the mountains "whose appearance", he wrote, "was of nothing more than a scabbed head" – the same heather clad mountains many people now enjoy as part of the Highland holiday view. Nonetheless, it still seems to be urban people who value the wilderness most, as, for example, the survey by Shafer and Hammitt (1995) showed, with visitors to wilderness areas being overwhelmingly from urban areas. Cronon (1995b) rather wryly observes, however, that these people who enjoy the recreation in the wilderness are the classes privileged enough to have the time and resources to put their jobs behind them for a while, rather than the less mobile and poorer groups in the cities or those who actually earn a living in or around the wilderness areas. Many urban dwellers flock to National Parks and similar areas at weekends or on vacation. This issue here then becomes one of the visitors actually detracting from the "wilderness" by their presence or by physical damage, often associated with trampling or vehicle use, while litter and other refuse may also become an issue.

The writer Thoreau (1854) wrote of the value of nature as an antidote to city life and of mystery and of forces greater than the self:

> We need the tonic of wildness, to wade sometimes in marshes where the bittern and the meadow-hen lurk, and hear the booming of the snipe; to smell the whispering sedge where only some wilder and more solitary fowl builds her nest, and the mink crawls with its belly close to the ground. At the same time that we are earnest to explore and learn all things, we require that all things be mysterious and un-explorable, that land and sea be infinitely wild, unsurveyed and unfathomed by us because unfathomable. We can never have enough of nature. We must be refreshed by the sight of inexhaustible vigor, vast and titanic features, the sea-coast with its wrecks, the wilderness with its living and its decaying trees, the thunder cloud, and the rain which lasts three weeks and produces freshets. We need to witness our own limits transgressed, and some life pasturing freely where we never wander.

Interestingly, in more contemporary writings, Bryson (1997) in *A Walk in the Woods* again betrays a dualistic attitude to wilderness, valued and threatened on the one hand and threatening on the other, but he is also rather ruthless with the romantic notions of Thoreau. Bryson sets off on the Appalachian trail "in a wilderness few have seen" deriving a pleasurable feeling of grittiness and determination to see "one of the world's great hardwood forests – a relic of the richest, most diverse sweep of woodland ever to grace the temperate world" which he saw as now under threat from global warming (p. 12). He then pro-ceeds to reveal how the woods were full of peril and wild animals especially. After some time in the woods he records (p. 66) how unnerving the forest could be. He follows Cronon (1995b) who recorded the trepidation that Thoreau felt in the forest, and indeed how Thoreau had felt diminished by the forest. Bryson writes that while the "inestimably priggish and tiresome" Thoreau "found nature splendid" this was "so long as he could stroll to town for cakes and barley wine" and that when Thoreau actually experienced real wilderness "he was unnerved to the core". The experience apparently left Thoreau, in the words of one biographer, "nearly hysterical". After some time in the woods, Bryson writes: "This wasn't the tame world of overgrown orchards and sun-dappled paths that passed for wilderness in Thoreau's Concorde, Massachusetts, but a forbidding, grim, oppressive, primeval place that was (and here he quotes Thoreau) " 'grim and wild … savage and dreary', fit only for 'men nearer of kin to the rocks and wild animals than we' ". He also records Daniel Boone's description as "so wild and horrid that is impossible to behold … without terror". Maybe now wilderness is still something we like to know is there but we only go into it on certain terms?

Wilderness is indeed both a romantic notion and hard place in which to survive – but also (and maybe even because of the latter), we look to it as a place for re-creation as much as for recreation. Combinations of romantic notions, taking your problems to the wilderness and the difficulties encountered abound in the literature. The are indeed many graphic writings about wilderness experiences and the difficulties of survival. Recently Krakauer (1998) in *Into the Wild* has written about the true story of "an idealistic 22 year old … who in 1992 walked deep into the Alaskan wilderness and whose SOS note and emaciated corpse were found four months later". Earlier, Jack London's (1910) *To Build a Fire* charts the vulnerability of a man in the snowy wilderness, the joy of lighting a fire and the despair when a snow laden branch drops its load and extinguishes the fire. The subsequent pages see the character go down through the layers of civilization, thinking of killing the dog and crawling into the warm carcass. In the end the man slowly dies and the dog runs off to find other fire providers.

The subject of Krakauer's book had family problems, turned his back on civilization and had indeed been inspired by Jack London's stories. The author writes how "people think the unsullied enormity of the Last Frontier will patch all the holes in their lives. The bush is an unforgiving place, however, that cares nothing for hope or longing". He talks not only of the personality type that can find fulfilment in comparative isolation: "a person whose principle need was to find some kind of meaning and order in life which was not entirely, or even chiefly, dependent upon interpersonal relationships" (quoted from Anthony Storr (1989) *Solitude: A return to the self*). But also Krakauer writes how many people find release through the nature of the wilderness experience: "The accumulated clutter of day to day existence – the lapses of conscience, the unpaid bills, the bungled opportunities, the dust under the couch, the inescapable prison of your genes – all of it is temporarily forgotten, crowded from your thoughts by the overpowering clarity of purpose and the seriousness of the task in hand". Also (when climbing) "you get used to rubbing shoulders with doom, you come to believe in the reliability of your hands and your feet and your head. You learn to trust your self control".

Such attributes are clearly worth developing, even though the author also admits that the problems do not actually go away but are often merely suspended, though they may be put into perspective: "I thought that climbing … would fix all that was wrong with my life. In the end, of course, it changed nothing." Wilderness is then perhaps now important as a concept of somewhere where we can go where human concerns seem petty and there is a sense of something "other" than ourselves.

This something "other" is seen by Bonnett (1996) as involving the relationships between identity, cultural appropriation and the landscape. Bonnett focuses on concepts of masculinity and quotes Harding (1992): "What is driving men in ever increasing numbers into the woods to beat drums and share grief that they can't articulate in their everyday lives?", noting that Robert Bly's (1990) *Iron John* topped the bestseller lists for 35 weeks in 1991. Central in Bly's book is the importance of men's ability to "free themselves from dependence on mother-figures and embrace their own autonomous masculinity". This has a parallel in the writings of Draper (1998) on the "soft intimacy of hard men in the wilderness" in South Africa, showing that in the wilderness there is an important home for men seeking to "shrug off their hegemonic power and attempting to move against their ingrained habits of their race and class identities". Bonnet's interpretation is that the rural, wilderness context facilitates "mythopoetic men ... creatively re-working colonialist fantasies of non-Western societies and landscapes", reflecting male powerlessness in a feminized world. This is expressed as a search for "origins and absolutes ... of unspoilt nature and uncontaminated humanity, for the paradise we (modern Westerners) have lost". The intellectual heritage is summarized as "wilderness philosophy" (Oeschlager, 1991) based on the articulations of Thoreau, Muir and Leopold, which seek to encourage "a spiritual and practical reverence for landscapes that are untouched by human hand and replete with the silent wisdom of authentic indigenous peoples". Thus, while there is much discussion of the role of rituals and how the movement to retreats in the wilderness may be regarded as part of a "male backlash" together with other interpretations, it is clear that "wilderness" is a construct where we seem to have a basic need of a place to go in times of personal insecurity where we can construct and affirm sets of values in a way that is not possible without it.

Schama (1996) in *Landscape and Memory* again relates the presumption, following Thoreau, that "wilderness was out there ... and that it would be the antidote for the poisons of an industrial society. But of course", Schama continues "the 'healing wilderness' was as much a product of culture's craving and culture's framing as any other imagined garden". "... the landscape that we suppose to be most free of our culture may turn out, on closer inspection, to be its product." "... it seems right to acknowledge that it is our shaping perception that makes the difference between raw matter and landscape." Schama talks evocatively of the "cultural backpacks we lug with us on the trail" and he quotes Ansel Adams: "Half Dome is just a piece of rock ... There is deep personal distillation of spirit and concept which moulds these earthly facts into some *transcendental emotional and spiritual experience.*"

Even more evocative is the story of Gilliatt marooned alone on a rocky island in Victor Hugo's (1866) *Les Travailleurs de la Mer*. It is said that Hugo wrote standing up to give him a sense of realism in his writings and certainly a graphic weariness of the struggle to survive imbues these writings. Eating his last biscuit, the central character Gilliatt is left without civilization and only the strengths of his own mind and body. He drinks from rain pools with the gulls and eats limpets off the rock. At one point ("Sub Umbra") Gilliatt wakes at night and peers into the indivisible darkness and there are thoughts of horror and terror hidden in the nothingness in what must be one of the best writings about the deep fears of mankind in the dark. "Darkness has unity, hence arises horror; at the same time it is complex, and hence terror." The text then, however, includes the words: "All is incomprehensible, *but nothing is unintelligible*" leading on to a section discussing the emergence of beliefs and faith in a way which tells us about the wilderness experience – of going back to some primitive state but thus also justifying the need for faith as some kind of way of ordering the universe. There is also the sense of feeling ourselves more – of sensing our own humanity when surrounded by an utter lack of human comforts: "His storehouse empty, his tools broken, tormented by hunger and thirst, his suffering left him wounds, ... bleeding feet, wasted limbs, pallid cheeks and eyes bright with a strange light; but this was the steady flame of his determination." Stripped of civilization, he had only his determination. While rather extreme and graphic, does this not indicate to us that this is why people seek the wilderness experience (albeit now often cocooned safely with technology and even maps)? To feel at least some slight sense of suffering in order to sense your self-esteem, will, determination and humanity? Perhaps, then, the wilderness experience is one of re-creation through finding that sense once more through physical experience and, again, the sense of something "other" than yourself.

The sense of "other" is also illumined by the writings of A.S. Byatt (1997) in *Unruly Times: Wordsworth and Coleridge in their time*. The author writes of Wordsworth's view of the defilement of nature "the increasing accumulation of man in cities ... minds blunted ... to a state of almost savage torpor" – Wordsworth hoped to rescue these minds in relation to the powers of mountains, sky and rivers; darkness and light – the desire for the unlimited and the uncontrolled – cataracts, mountains, winds, avalanches. The fundamental explanation for this attraction to wilderness is given in terms of the writings of Burke in 1756 on the Sublime and the Beautiful. Byatt tells how Burke presented an analysis of *why* terrifying, infinite or empty things were attractive to the human mind through the division of sentiments into the beautiful and the sublime, the latter being fundamental:

> **Beautiful** – important qualities: soft, smooth, harmony and appeal to the instinct of self-propagation.
> **Sublime** – appealing to the human apprehension of fear, pain, thwarting forces or infinity – concerned with the instinct of self-preservation.

Attributes of the sublime are seen as:

1. Obscurity (which induced terror and as seen in Milton's darkness).
2. Power.
3. Privations – vacuity, darkness, solitude, silence.
4. Vastness – height and depth.
5. Infinity – a tendency to fill the mind with that sort of delightful horror and artificial infinity which involved:
6. Succession – an endless progression beyond limits; and
7. Uniformity – a round and therefore artificially endless church, the pantheon as opposed to the cruciform church.

The sublime is seen as *a sense of real power or danger faced without damage*:

> When without danger we are conversant with terrible objects – the mind always claiming to itself some of the dignity or importance of the things which it contemplates.

This, then, to me, increases our fundamental understanding of the wilderness experience. The key elements are that the awe which is inspired by the scenery transfers some dignity to our minds while the secondary element is one of putative danger but no great risk.

Re-creation, then, in its true sense, can indeed come from "the wilderness experience" (if you are prepared for it enough so that you come back!). In his book *Nature Recreation: Group guidance for the out-of-doors*, Vinal (1940) wrote of an "armistice with the city" and storing up "a good reserve of antidisease, unadulterated sunshine and uncontaminated sea breeze or forest air" for young people. A more modern approach is covered in Hammitt and Cole (1998) in *Wilderness Recreation: Ecology and management*. There are also many later accounts of the therapeutic value of wilderness in the literature on human psychology, especially when some traumatic experience has been suffered (for example, Asher (1994) wrote of wilderness therapy for rape survivors, Angell (1994) writes of the wilderness solo as an "empowering growth experience for women" and Powch (1994) and Cole *et al.* (1994) write of "Wilderness therapy for women – and its empowerment").

There are many other accounts of the importance of wilderness to the human psyche, from the generalized points made by Laurens van der Post to individual accounts – and it is striking that many individuals undergoing wilderness experiences as re-creation do so alone and relate their experience in a book as a witness.

Laurens van der Post (1986) in *A Walk with a White Bushman* talks of a conservationist friend:

> ... [he] was struck by the fact that conservation is a deep need in the soul of man, ... man needed it almost more than the animals did, and that once he brought modern people, so estranged from nature, back into contact with animals in their natural wilderness surroundings, people were changed by it for good ... so our whole aim is to re-create, or re-educate, individuals through nature and to see life from nature's point of view, so that wherever they go they take this vision with them.

> We need as many [wilderness sanctuaries] as possible. Every bit of unspoilt nature which is left, every bit of park, every bit of earth still spare, should be declared a wilderness area as a blueprint of what life was originally intended to be, to remind us. When we do see that, it is like having a religious experience, we are changed by it. This act of instinctive remembrance sparks off a most dynamic sense of guilt and horror at what we are doing, and an immense longing not just to conserve but to rehabilitate the earth.

> Once man feels himself part of nature again instead of just exploiting nature – it is only man who has put himself outside nature – and once man allies himself to the evolutionary energies and forces of nature, there is nothing he cannot change, including himself – which is the hardest thing of all. And this lesson you learn in wilderness very quickly.

... nature will always take care of itself even without man. But man without nature is unthinkable and, known or unknown, his spirit needs it: needs it for its survival, sanity and increase, as his body does. The real things in life have nothing to do with the question "why", they are just "so", they are just "thus". Life is a "thus", and until you realise this "thusness" of life you are stuck.

In terms of a personal experience, and while the Lake District in the UK may or may not be regarded as a wilderness, the experience-base is similar, Wyatt (1973) wrote *The Shining Levels: The story of a man who went back to Nature*. He writes of a pointless existence, a meaninglessness and a loss of identity in city life and of his deep need for contact with nature:

[After leaving the Navy] I went back to the old urban charade ... Then it came to me – as a sort of sickness. One morning I saw myself, not as myself, but like a character in a play. There I was: a man with a name which meant nothing. The face I saw reflected in the mirror was that of a stranger. The body I wore was not mine. I had lost my way.

For some days I did not know what was wrong. Then I knew. I had to go back.

The Lake District to me was not the soft, mild benevolent, romantic place that the holidaymaker knows; was not the wide free fells or the challenging crags. It was, as I remembered, the smell of moist earth and newly cut wood; an aching back; rain in my face and wet feet; blisters on my hands; sore muscles; hard work, and a bitterness as well as deep and total satisfaction. The Lake District was my hard taskmaster, my teacher, my spiritual guide; never my mistress. I had to return.

The author then goes to live in the woods and the rest of the book tells the story of his experiences with nature.

What is interesting to me in this section is that in writing about wilderness we have touched upon contact with a non-human world and experiencing a sense of human nothingness with the concomitant enhanced awareness of the need for belief, faith, science. These needs, of course, vary with different people – and some are deeply content to live in cities – but they seem to be deep within us all somewhere. And surely these senses should inform our debate about responses to environmental change, and discussing them is important. Such senses are probably more fundamental than research on biodiversity, studies of plant responses to elevated carbon dioxide or predictions of increased temperature. The *significance of these researches lies in indicating what changes might occur and then presenting challenges in terms of what we value; values are therefore seen as more fundamental.* Countryside, parks, gardens and nature all provide contact with nature, but to a significant group in society, to be in touch with nature gives you a sense of contact with something other than yourself. This seems to me to be fundamental to but also larger and deeper than the climatic change debate. This much is written large in the wilderness experience.

What is then interesting is the way that wilderness protection often expresses itself not so much through the psychology of "untouched" spaces but through ecological, conservation, rarity or some scientific construct – rather than as it is,

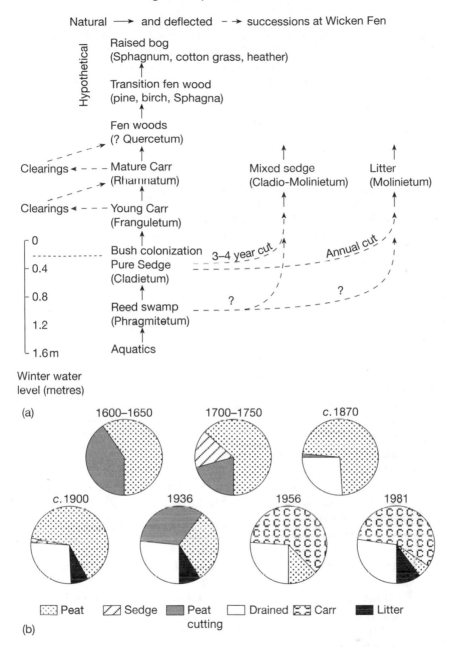

Figure 5.2 Wicken Fen. (a) The left hand sequence charts the succession from an open aquatic habitat through the process of terrestrialization. The mixed sedge and litter communities are only maintained by the human intervention of cutting. (b) Land cover 1600–1981. The litter and sedge communities dominate because of cutting management which declined in the early 1900s. Carr takes over without cutting and sedge and litter habitats are now only returning with renewed cutting. It is perhaps ironic that the habitat seen as desirable for butterflies, other insects and birds is only maintained by human intervention (from Friday, 1997).

national possession and heritage, they belong to nobody because they belong to us all; nor do they belong exclusively to us, because the generations that come after us have exactly the same right to the enjoyment of their beauty as we have. Man did not create these other sharers in the mystery of life, and man, unless they threaten his security or existence has not the smallest justification in effacing them from the world.

Following this statement of the sanctity of wildlife, there is, however, the notion of human-centred control and creation:

And the maker of a sanctuary is writing his own nature poem in his own way; or we may put it that his is the little-great part of healing nature of the wounds of greed and vandalism of men have inflicted upon her. Nature in return admits him to a knowledge of her secret ways which will brighten and refresh his spirit and afford it a strength and consolation which he will learn to value with those permanent elements of beauty, peace and human affection that make life worth living.

The spiritual importance of nature reserves thus seems to involve both preservation and active involvement and intervention. Nature reserves are therefore putatively natural but preference and choice are guiding principles in management. In many cases diversity may be a preference and/or key species seen as particularly desirable. Rarity has an especial cachet but representativeness is also important. In most cases there is a management plan (e.g. Usher, 1973; Sutherland and Hill, 1995), indicating the intervention which is necessary to achieve the desired state. In this context, enormous efforts are often expended in terms of controlling vegetation succession and also in preserving oddities which may often be relics of former conditions or may be otherwise fragile and liable to loss. The decision to do nothing is actually a positive decision, indicating that the current status is a desired one and either not liable to change or that current progressions are also desirable.

Tension may exist between those who espouse a "naturalness" and those opinions, often "expert" which promote the cause of one or other species or habitat. As mentioned in Chapter 1, Harrison and Burgess (1994), in their study of people's attitudes to conservation concluded that while the general public have a "common sense understanding of how nature is organised" (see also Eden (1996) on the public understanding of science), nature reserves were often managed in terms of "a rhetoric that extols the intrinsic worth of organisms" and that this is "based on the assumed universal authority of scientific 'experts'". This may well lead to a sense of lack of involvement by the public in setting conservation priorities. When they are involved, there is often also a nostalgic element where the criteria for management may relate to some previous state seen as desirable, often in relation to the memories of older people who retain images of conditions when they were younger – and when (all) things were putatively "better".

The history of British nature conservation is recorded by Evans (1997). Many of the formal societies involved data from the period between 1820 and 1890. The desire was mainly one of protection and, for example, the Selbourne Society for the Protection of Birds, Plants and Pleasant Places was formed in

1885 with aims which included the "preservation from unnecessary destruction of such wild birds, animals and plants as are harmless, beautiful or rare" and to "protect places and objects of interest or natural beauty from ill-treatment or destruction" together with a discouragement to wearing bird plumage as ornament (unless the birds were killed for food or reared for their plumage) and also an aim to promote the study of natural history. Many other markers were laid down around this time mostly aimed at some kind of protection, especially protection against development. The Royal Society for the Protection of Birds (RSPB) grew from the Society for the Protection of Birds (1891) and was granted its Royal Charter in 1904, the National Trust formed in 1895. The Society for the Promotion of Nature Reserves (SPNR) was formed in 1912, largely at the instigation of Nathaniel Rothschild. Its aims included the notions of "retaining primitive conditions" and, the press coverage of the time included the telling headline: "Menace of the Towns: Saving Wild Life" (Rothschild and Marren, 1997).

A useful review of ecology and the management of nature is provided by Adams (1997) and both he and Evans (1997) refer to the foundational committee, chaired by J.S. Huxley, reporting in 1947 in *Conservation of nature in England and Wales*. Here still the key words were *preserve* and *maintain* but also the duties of the new national Nature Reserves would include not only the selection and acquisition of nature reserves, with survey, research and educational work but also the word *manage* was used. Evans then writes of refuges and reservoirs, together with the threats from pollution in the 1950s and 1960s with the British Butterfly Conservation Society, formed in 1968, with its aims to "save from extinction or protect". The Woodland Trust, formed in 1972, again talked of retention but also used the word *enhance* with regard to sites as landscape features. The SPNR became the Society for the Promotion of Nature Conservation (SPNC) in 1976, again talking of reservoirs for rare species and safeguarding scarce habitats. By 1978 Fitter and Scott (quoted from Evans (1977)) were still talking of unspoiled ecosystems but Mabey (1980) was also talking of the needs of the wider countryside, regarding "site protection ... as no longer adequate or very appropriate (as an) approach to the wider countryside". And Evans then quotes HRH Prince of Wales as saying in 1984 that "there is a danger of assuming that the objects of conservation can be fulfilled simply by creating reserves ... (what is needed is) a conscious and continual effort".

A milepost on the transition from protection to active management was perhaps the formation of the British Trust for Conservation Volunteers (BTCV) in 1959 for practical management work – often involving scrub clearance and footpath work. Management thus increasingly appeared as a process, management plans becoming a central requirement of English Nature (formerly the Nature Conservancy Council) at the present day (see Inset 5.1). In 1995, Sutherland and Hill published *Managing Habitats for Conservation* on practical management, including the all-important management plan with aims and a discussion of the principles of ecological management on which they might be based. The telling sentence (p. 5) is "many of the most *precious* [my italics]

Inset 5.1   Conservation objectives in Slapton Ley National Nature Reserve, S. Devon (Trudgill et al., 1996)

Guiding Constructs:

Habitats and diversity
Naturalness
Rarity
Fragility

Specific objectives:

To maintain the aquatic habitats – water level and water quality

To maintain and enhance the populations of rare species

To maintain, enhance and protect existing diversity of habitats and vegetation:

>   Shingle ridge – scrub clearance
>   Wetland – extent and quality of reed beds
>   Semi-natural woodland – landscape and coppice
>   Grassland – wet meadows

To control undesirable or pest species

Seek additional protection on areas adjoining the reserve by liaison/ agreement with landowners

Legal: South Hams District Council, boundary/trespass, Public Rights of Way, good relations with neighbours, visitors, authorities

Provision: site monitoring for effective reserve management, scientific research and survey, education and interpretation, angling, controlled access

wildlife communities in Britain are at an early successional stage and thus much of the effort of conservationists is in suspending succession", followed by "The intensity of management is inversely proportional to the maturity of the habitat being managed". There is, however, no definition of "precious" but implicitly in the book this involves rarity, together with representativeness and diversity.

Sutherland (1998) in *Conservation Science and Action* does have a discussion of biodiversity as a social construct but again we have the statement that "A major reason for the need for intervention is that many species *of conservation importance* [my italics] are dependent upon uncommon, early successional habits which need to be maintained at an early ecological stage by cutting, grazing, burning or disturbing (Sutherland, 1998, p. 203) but we are not there told what "conservation importance" is. The author also notes, however, that there is now a growing call for a decrease in human intervention and increasing interest in restoring natural processes (p. 202). Does this go back to letting nature take its course?

Adams (1997) reviewed the emerging need for management, citing the example of Woodwalton Fen in Cambridgeshire. Here, as at Wicken Fen, the site was about 90% covered with birch or sallow scrub in 1959 when it was leased to the NCC. Again, controlling succession was the objective. Usefully, Adams distinguishes between the cultural values attached to nature and the rationalist project of conservation. Sutherland and Hill (1995), in stressing the importance of the management plan, provided a summary of the rationalist approach. Here the information has to be collated, the objectives set, the necessary organizational procedures to achieve the objectives and the need for monitoring the effectiveness. Such an approach may be based on the identification of the prerequisites for the preservation of species such as is provided by Pimm (1991). Pimm sees the following points:

- Population size is important as this increases the time a population lasts in the face of demographic accidents (for example temporary decreases).
- Population size is of far less importance where environmental disturbance is involved – large population size may then offer little or no protection against extinction.
- Slowly growing populations are more vulnerable because they recover less quickly from environmental disturbance.

Such statements are a useful guide in our eternal search for answers to the questions – *what do we do for the best – how should we act – what are we trying to achieve – what should we be guided by?* In a mechanical, rationalist sense it means that population sizes should be enhanced, environmental disturbance minimized, that slower growing populations might need more protection, and so on. Cultural values, however, underpin all these kinds of statement – especially that a species is, in fact, desirable. This takes us to the realm of *meaning* which Oliver Rackham (1986) describes so well (see Plate 5.1). He especially places value on the historical context of the landscapes, drawing parallels with the irreplaceability of unique old paintings.

Significance and meaning are thus in many ways derived from association and meaning also varies with cultural context. Historical associations can certainly be a powerful factor in the perception of significance. Other meanings may simply be ascribed to beauty and a sense of awe and wonder. In particular, though, if you watch how people behave on nature reserves and listen to what they are talking about, it is clear that much excitement is generated by rarity and unusualness as much as by beauty. *Rarity is a relative associative construct, meaning and significance decreasing with a rise in numbers* (where ospreys are commonly to be seen, as in parts of North America and Scandinavia, for example, they do not excite as much interest as the few that started to nest in Scotland for the first time some years ago). A sighting of a rare bird or plant also carries with it an associated notion of *privilege* – not everyone or anyone has seen or can see this sight. Often this involves a chance sighting with associated notions of happenstance or serendipity as, for example, when a North American bird is blown off course on migration and seen in the UK. "Stray" visitors may stir more excitement but are usually seen as less significant than birds which nest

Plate 5.1 Hayley Wood, Cambridgeshire. An ancient woodland. These are seen as irreplaceable but have been used intensively by people in the past for wood resources. Their meaning lies as much in the history of their past use as in any sense of "naturalness".

here for the first time. Similarly unusual habitat fragments and remnants seem to evoke a positive response. Additionally, the speciality of a particular place is often involved. Indeed Sutherland (1998) exhorts managers to conserve what they have got and are "good at" – do not dig a pond in a grassland just for the sake of diversity – conserve the grassland because that is your speciality, especially if your grassland is one of the few cases in an area. This means that there is an *associative significance placed on a site in terms of typicality and unusualness.*

Given that we do live in worlds of significance and meaning, the book by Perlman and Adelson (1997), *Biodiversity: Exploring values and priorities in conservation*, is important in that it faces conceptual and perceptual issues squarely. With the implicit question of "Why is biodiversity important?" (i.e. what is its significance and meaning?), and as we saw in Chapter 3, they pose the question: "If biodiversity is good, is more biodiversity always better than less biodiversity?" (p. 23). Their reply rests on a case study of an introduced species, where the answer tends to be a simple "no" because integrity of the "original" ecosystem also seems to be important. They then proceed to the question of "Does more different mean more important" and arriving at the conclusion that there is no one right answer, they propose that the worth of an element of bio-diversity depends upon the context, and in a discussion of particulars, contexts, values and worth come squarely back to the fundamentals of human values. They then sideline the numerical view of biodiversity by referring to the more important system of values (p. 39). Value systems include the motives, prefer-ences and underlying belief system that a person has in undertaking an activity, investigating a matter or in protecting an object and also alternatively as a more economic evaluation including use value, option value and existence value: "Worth is seen as assigned when an object satisfies or in some way matches one of the values held by an individual". Worth, then, thus depends upon the values that are important to individuals. The example they give is that if cleanliness is valued by an individual, then detergents have great worth (enabling your value).

Perlman and Adelson tend to legitimate all value systems and encourage the listening to all of the "voices" involved, citing Kellert's (1993) classification of the human values of nature which is reproduced here.

Type of value	Definition
● Utilitarian	Practical and material exploitation of nature
● Naturalistic	Satisfaction from direct experience/contact with nature
● Ecologistic–scientific	Systematic study of structure, function and relationship in nature
● Aesthetic	Physical appeal and the beauty of nature
● Symbolic	Use of nature for metaphorical expression, language, expressive thought
● Humanistic	Strong affection, emotional attachment, "love" for nature
● Moralistic	Strong affinity, spiritual reverence, ethical concern for nature
● Dominionistic	Mastery, physical control, dominance of nature
● Negativistic	Fear, aversion, alienation from nature

One might add *indifference* to this list but it can be seen that worth can be expressed in terms of any one of these value systems. As Adams (1997) points out, the "ecologic–scientific" value system has perhaps tended to dominate nature reserve management in the past. This is blended with the moralistic value system to value rarity, fragility and remnant features as well as representative-ness. Reviews of the importance of existing protected nature reserves relative to areas which might also be considered as worthy of protection frequently include assessments of representativeness/typicality and of rarity. In addition, the control of succession is not without the dominionistic value system – as such it is almost a form of gardening and certainly an expression of preference.

This is all of considerable interest if we are faced with the prospect of environmental change and the attendant challenge to our value systems. At the very least it may mean that the protection of some relict plant communities may become more difficult, especially, say, patches of arctic–alpine plants, relict from former colder climates, in a warmer world. What is interesting is that we are trying to preserve particular states at particular sites. This usually involves the manipulation of states rather than processes and may involve a simplistic view of complex systems and their internal interactions. It might even assume that there is a knowledge of the relationships between causes (for example, inputs to a system), intervening processes and effects (states) and the assumption that manipulation of the apparent causes will lead to desired effect.

Ecosystems are, however, complex, as we have discussed in Chapters 2 and 3. Linear extrapolations are not feasible and the outcome of actions are not wholly predictable (Pahl-Wostl, 1995). This system complexity means that whatever action is taken can have unforeseen effects. There may thus be wider consequences beyond the intended target of the action and the proposed target for the action might not actually benefit. An example here might be that taking action to decrease river pollution and sediment to encourage fish might have the consequence of clearer waters making the fish more visible to predators leading to increased predation and a drop in fish numbers. Alternatively, we might think that we have established the chain of cause and effect but there are also other factors involved. Here, for example, decreasing nutrient input to eutrophic lakes may not lead to a decrease in algal growth as this still might be facilitated by the lack of herbivorous zooplankton, the numbers of which are actually minimized by inputs of pesticides. This situation makes single-aim conservation much easier than anything else. A simple case is when we clear scrubland to re-create grassland for butterflies, for example, where we are expressing a preference for a state of increased numbers of butterflies. Success is readily judged by counting the numbers of species of butterflies. Creating the right conditions for one species, such as the Osprey in Scotland, focuses on one species, a set of habitat, nest site availability and feeding conditions together with controlling human access. Such focused management involves one set of actions and one desired effect and may purposefully ignore the effects on other ecosystem components. Thus other things might happen but they are seen as of lesser significance and marginalized – the criteria of success are obvious and every effort is focused on the target, with re-thinking when the desired effect is not achieved being about improving the strategy rather than altering the aim. Complexity of the system is thus of lesser importance in single-purpose conservation – the attempt is refined until the purpose is achieved (such as, for the examples given above additionally providing plant cover for fish or reducing pesticide use in a catchment above). Multi-agenda preferences are much more difficult to accommodate as the effects of achieving one aim might be seen as undesirable for achieving another. Indeed preferences, criteria and proposed actions may be contested, especially if they clash. Here one of the main roles of the biodiversity concept is that at least this is one thing that all can agree on, with the fact that multiple interpretations of the term are allowed, perhaps having merit rather than being a failing (Perlman and Adelson, 1997, p. 11). Diversity then becomes the aim and *the facilitation of multiple meanings becomes the criteria by which success is judged.*

A multiplicity of meanings is not necessarily related to a multiplicity of states – one state can have many meanings. What this then means is that it might be preferable not to actually attempt to specify and accommodate a multiplicity of preferred states. This attempt might lead to almost endless intervention in order to allow for all the side effects and continued "corrections" to allow for all the different agendas. I have seen this on nature reserve management committees where actions to enhance conditions for one species are seen as detrimental to other species, for example some conditions might favour some species (clearing reeds to provide open water) but be detrimental to others (reed-nesting and feeding birds, and as was also discussed for Wicken Fen and water conditions, above). The arguments might be resolved by the higher claims of one lobby in terms of rarity (locally, or better, nationally) or we may even have to accept that the outcome may be related to a stronger and more vociferous lobby on behalf of one species which relates more to the personalities on the committee rather than to anything else. In addition, specialist biologists and even more broad-based ecologists might perceive the whole process as more difficult if demo-cratization is involved as the general public tend to have a preference based on the naturalistic, aesthetic and humanistic sets of values, just liking trees, flowers and birds in general. This tends to pit a more elitist "speciesism" against what could be regarded as a less discriminatory democratization. This in turn poses the question of "nature conservation for whom?", making the outcome cultur-ally and socially positioned. This may be seen as entirely fine, but the outcome is likely to be one of the balance of different "voices" and "small voices" may not be empowered despite efforts for participation.

Earlier in this book, we proposed that some essential features of ecosystems were:

1. Heritage – left from past conditions (which may no longer exist).
2. Renewal – the processes which ensured future states.

The proposal indicated that states are renewable but not repeatable. We may now clarify the dual purposes of nature conservation.

1. Intervention to preserve a preferred state – to prevent change and to prevent the loss of heritage (which can be irrespective of whether the conditions maintaining that state still exist or not, though it is clearly more difficult if they do not).
2. Intervention to facilitate change – enabling species to continue and react to new conditions.

These aims are not mutually exclusive and the former has characterized much of conservation hitherto while the latter, it will be argued, is an important consideration for the future.

Intervention to prevent change

If some state is intrinsically desired or preferable, often in relation to the moralistic value system of nature, one is talking simply of preservation. Here, it is the historical legacy we are trying to keep, much as one might preserve a Norman castle or cathedral: the Norman era can never be repeated and we give a Norman castle worth in terms of this fact and see it as significant and meaningful as part of our heritage. This is the approach advocated by Oliver Rackham, cited above. It can be extended to cover relict patches of habitat, such as the arctic–alpine communities in Scotland, and allows for a nostalgic approach and refers to the moralistic attributes of reverence and ethics.

With climatic or other environmental change you are then presented with the same dilemma as those who have charge of deteriorating old buildings. Building stone which has been weathered over time has value because it is old and original – and the origins cannot be repeated – and even the weathered state is valued, giving it a venerable significance, even as a ruin. The dilemma is that loss of statuary, carvings, and so forth, can be seen as disfiguring and it may be that if nothing is done, the building could collapse. If one accepts this latter kind of judgement, the only alternative is to renew the carving and place fresh stonework in its place. This preserves the visual integrity of the building and can be justified in terms of the building being a "living" evolving building but detracts form the historical significance. This all causes a deep examination of values and attitudes to relics – either preservation or renewal may easily win the day.

The same kind of dilemma is evident in conservation of relict habitats or species. Here, however, one has to ask how far the conditions which give rise to the preferred state still exist. Unlike Norman castles and cathedrals, however, ecosystems have the possibility of self-renewal. If the conditions that are appropriate to a particular preferred ecosystem are maintained through time, then the management tasks are simply to make sure that these conditions are protected and continued and that, usually as a consequence, no other state occurs other than the preferred one. If the conditions are not maintained, then this means that with relic ecosystems either we preserve, maintain, enhance or re-create the conditions which are appropriate to the relic or become reconciled to the loss.

This "heritage" approach also justifies the removal of succession to a large extent, though there will still be arguments over the time reference of the preferred state. The answer may be one of "within living memory" or of maximizing historical value. In Wicken Fen, for example, a variety of vegetation covers have existed in the past (see Figure 5.2). Currently, the management plan refers back to creating sedge and hay with reference to the fact that there is interest in the fen as a resource used by people for products as much as there is current ecological interest in diversity. Historical interest, is, however, the guiding principle. This historic interest is independent of any considerations of biodiversity, species or habitat preferences but is not opposable to them and can operate in tandem, especially as rarity of historical occurrence, rarity of species and diversity may often (but not necessarily) operate together. In turn, rarity may often imply a fringe of distribution which itself may have a context in terms of the historical evolution of vegetation.

The excitement over rarity and historical respect may thus converge. Underlying this is a cultural attitude of the value of history and an interest in rarity and curiosity which involve values which are widely shared. Proposals for management actions to preserve objects of historical significance and/or rarity can thus be readily democratized widely. The proposals can be discussed and decisions can then be made from a broad, participatory basis in a society which shares the same cultural values – and they do not necessarily require any specialized or "elitist" knowledge. In terms of species "in danger" Bendall (1996) rehearses the criteria as involving decline in numbers or range, endemism or species under a high degree of threat as internationally recognized. The heritage approach is thus culturally defined and independent of ecosystem science (with its associated uncertainties rehearsed in Chapter 3, except in terms of an all important understanding about the conditions which will sustain the preferred state). It does, however, require a consensus about the preferred state. If maintaining the conditions becomes untenable, there may also be a need for a possible revision of the preferred state over time, though the very nature of this purpose is to preserve the past rather than to update the preference. The criteria of success lies simply in the continuation through time of the preferred state.

Intervention to facilitate change

This second purpose of nature conservation lies not in specifying states and attempting to rationalize the agendas of specialists or lobbies, nor even in the expression of preferences. It certainly is not about manipulating each component of the ecosystem to a preferred state (ecosystems are far too complex and mutually reactive for that, as discussed above). It does not even necessarily lie in a striving for biodiversity. Despite all the sophisticated arguments for biodiversity (Perlman and Adelson, 1997) with associated statements about utilitarian values and moralistic arguments and contested associations with stability, biodiversity is simply a preferred state. Diversity is intrinsically more interesting to human beings than monotony. What facilitating for change actually lies in is the preservation of the processes of renewal.

This is not the same as allowing nature to take its course, though Sutherland (1998) does rehearse recent arguments for decrease in human intervention. It is about removing the conditions which prevent naturally occurring processes to happen. Adams (1996) expresses it as "*we need to allow nature to function, and create conditions for it to do so*". I am here expressing it the other way round in terms of *removing impediments*. This involves removing, say, *human induced* toxicity (bearing in mind, for example, that lead veins do occur in rocks and that plants are adapted to them), high nutrient loadings to lakes, pesticide inputs to rivers and so on. The reference point is about *minimizing human-induced conditions which hinder renewal*. It is not a romantic view of "nature" and "natural processes", more a study of how ecosystems – plants, animals;

species and habitats, actually renew themselves. It admits that preferred states might only be a temporary statement and that future generations might have other preferences. Central is the theme of preserving the possibility of future states (which may indeed differ from current ones).

This point, however, should be tempered by the point made by Hoffman and Parsons (1997) in *Extreme Environmental Change and Evolution*. While, under global warming, there might be direct temperature effects, there could also be a host of indirect effects and many interactions between food, predators, pests and parasites. The author points out that many pests and pathogens are culled by cold winters and if these decrease in number or severity, the pests and pathogens could spread their range. Such a spread could happen anyway, irrespective of human intervention, but intervention to facilitate the migration of (desirable) species to new ranges could also facilitate spreads of pests and pathogens.

Sutherland (1998) also rehearses the arguments for intervention – mainly halting the progression of succession as a means of preserving those species of conservation importance as flourishing at an early stage of succession. I think we may now say what this conservation interest might be. I would see this primarily as historical interest outlined above. We are trying to preserve something as it was and not permitting change. This is not, to my mind, particularly justifiable unless one admits that it has the cultural value "heritage" justification (or an educational one to demonstrate the stages of succession). It is simply preserving the past – and needing legitimization under the heading of heritage purpose – as, indeed, is biodiversity conservation (a range of species is part of our heritage). It is a cultural activity and should be seen as such. We must also realize that in many cases we are preserving a cultural artefact rather than nature (as, for example, is the case with coppicing).

Nature conservation should perhaps be re-named *species conservation, habitat conservation, diversity conservation* or even *preferred state conservation* because it does not seem to be much about conserving nature, that is leaving nature to get on with itself. This much is evident from the quotations we started this discussion with where, at Wicken Fen, allowing nature to occur was seen as undesirable. Intervention to achieve preferred states is the order of the day and almost involves a conceit about preferences rather than a humble approach to nature. Letting nature happen is certainly something entirely different and something almost of an anathema to the whole of the conservation movement. (The few contexts in which you can read of succession as a "good thing" come in terms of environmental restoration where Gilbert and Anderson (1998) write of *promoting* natural succession on derelict sites or in the notable book by Jordan (1998), *Working with Nature*, where a section (p. 94) on working with succession in order to provide a wider range of ecosystem products.) Indeed letting nature happen is something that we are manifestly not very good at – we are too much of a gardener at heart (Walton and Bridgewater (1996) justify the perspective that the whole biosphere is a "garden" and that people are gardeners), and despite expressions of a preference for wilderness – we always seem to have the need to achieve something, to leave our mark.

The argument is, then, that we are admitting that much of nature conservation is about maintenance of (and arguments about) preferred states (which include considerations of both past relics and of diversity). This is in relation to a number of different value systems but we have now managed to identify two separate purposes – *preserving the unrepeatable legacy* and *preserving the processes of renewal*. In the latter, disturbances might become the norm (Fiedler *et al.*, 1997). We might also recognize that restoration may not be a legitimate means of compensating for other losses (Eden *et al.*, 1999). What, then, does protecting the processes of renewal mean in practice? This is also as much a question from those undertaking environmental restoration as much as for those who undertake nature conservation (see Mace *et al.* (1998) *Conservation in a Changing World*). In essence it admits that change is normal and irrespective of current scenarios of climatic change, change is always likely. This is not to say that arresting succession is not legitimate (it is part of our expression of preferences) so the clearing of scrub and so on is not being denigrated. In fact it is, to my mind, firstly, not necessarily anything positive in terms of action, apart from activities like conserving seed types and other forms of reproductive possibilities, nor about preserving particular habitats, though keeping a range is clearly keeping open the widest number of possibilities. It is, as we said earlier, decreasing the conditions which are inimicable to life. Secondly, and beyond that, it is about studying the processes by which plants and animals disperse and change their range in response to changing conditions.

In this context, studies of dispersal, evolution and adaptation are crucial. This is especially so since we observed in Chapter 4 that from the study of Webb (1986) the conclusion was that: "*dispersal leading to a change in spatial range has in the past appeared to be a relatively rapid process and, indeed, the main way of responding to change.*" Central to this are considerations like corridors for migration, rates of dispersal and so on. Allowing nature to happen means more to do with providing opportunities and ensuring mechanisms for plants and animals to change their geographic range. This is not so much a cultural activity, concerned with preferred states, but about enabling change to occur. Earlier, we admitted that conditions which are appropriate to a preferred state might not be maintained; culturally, we may like that state, relict though it may be. We may then try to maintain appropriate conditions at that site. But this endeavour is not about sites, it is about how the species might have an opportunity to change its range, relocating at a different, more appropriate site. We can, of course, translocate species ourselves and this may be a viable option, especially where we are more sentimentally attached to more charismatic species. However, we cannot do this for all species and there is no particular reason why all should survive in the face of environmental change – the rule of the past has been one of environmental change, evolution, adaptation and extinction Given, however, that the rates of change may now be seen as perhaps faster than ever before (though this is debatable), the issue of habitat fragmentation and corridors for dispersal and migration become fundamental (see, for example, Corbit *et al.* (1999) for a discussion of the role of hedges as habitat corridors). This is then more a matter of facilitating nature to react – the control element lies not so much in manipulating species and habits but allowing reactions to change to take place. That is what letting nature happen means in the context of environmental change. It means a shift in priorities from intervention to prevent change to intervention to allow change to occur.

While this may mean changed priorities for each local nature reserve, it is far wider than discussions about succession in nature reserves and indeed a concern far wider than being about nature reserves at all. It is about all nature and all landscapes. Nature reserves after all only occupy a small percentage of the land surface in most countries. In particular, corridors for dispersal become important and considering habitat distributions in terms of connectivity.

It is clear that past environmental changes have lead to a variety of plant responses (Huntley *et al.*, 1995) including evolution and migration. It seems that evolutionary processes are, however, often too slow to cope with climatic change fluctuations and that migration or other changes in geographical distribution have been critical to survival in the past. However, this does mean that in order to facilitate such geographical changes now we must consider the fragmented and disconnected nature of habitat patches which exist in many areas now (Lindenmayer *et al.*, 1999). Collingham and Huntley (2000) have recently considered the optimum habitat patterns which may facilitate migration in a fragmented habitat where a mosaic of linkages emerges as a key point. Much wild life habitat is now surrounded by agricultural land which can hinder migration, though this is less of a problem for flying insects, birds and plants with wind-borne seeds than for some other plants. The questions of recruitment from nearby areas and mechanisms of dispersal become important here. In terms of actions on the ground, habitat preference could still be expressed by manipulations such as scrub clearance but it should also mean that planting should be aimed at enhancing and protecting linkages between sites as much as activities within protected sites.

In terms of philosophy, there is already a growing theme of enhancing/allowing nature rather than control. Adams (1997) argues that while the appeal of nature conservation has been emotional but in practice much of it has been in a "rationalist" context in terms of predicting and controlling nature. He quotes Nicholson (1957) as encapsulating this in terms of "management by interference" and also writes of "ideas about ecosystem succession (yielding) an explanation of the capacity of nature to change in undesirable ways". Also, he writes that "human action upset the machine" of nature and the ecologist, "spanner in hand", is needed to put the balance right. However, he feels that those notions of balance are no longer appropriate for the guidance of management action and cites Holland and Rawles (1993) as suggesting that nature conservation is about change: "*negotiating the transition from past to future in such a way as to secure the transfer of maximum significance*" which is very much in line with the discussion above. An interesting definition of conservation that still has resonance today is that of Westhoff (1983) "A social mechanism of feedback ... directed particularly at the relation of man with his natural environment". The idea of feedback between society and conservation is an important one.

Pahl-Wostl (1995, p. 208) expresses the clearest statements of this kind of philosophy "Living systems obviously have the potential to create new types of responses to unprecedented situations. Should we not explicitly include this property in our considerations?" and the author also follows Hollick (1993) in that "*managers should cultivate the capacity for (ecosystem) self-organization rather than trying to control them*", meaning that management should be "flexible and decentralised" (p. 213).

This again is not about maintaining preferred states but about conserving the processes of renewal and indeed the quality of "resilience". We argued in Chapter 3 that the *possession of the capacity for flexible responses to changed conditions* is a useful and appropriate way to view nature and that it may then be argued that *enabling a flexible response* should be a priority for management. The critical feature in nature conservation should thus be about *enhancing the ability to adapt to change* rather than intervention to maintain preferred states, though the latter is not excluded (out of historical interest, as discussed above). Capacity building and resilience building are thus key elements of this endeavour. In this context, the differential role of species (Holling *et al.*, 1996) with "drivers" and "passengers", as discussed in Chapter 3, and the identification of the "drivers" will be important here. This also emphasizes a key element in the justifications of biodiversity (Perlman and Adelson, 1997) but not in the context of "all species" but of "key species", nor in the context of "preserving stability" as in the older rationalist approach rehearsed by Adams but in terms of enhancing the range of chances of some species somewhere as being able to cope with the changes. Preference for particular species may still be evident and can be allowed for when it comes down to detailed management, but the philosophical stance is quite different from controlling for preferred states.

In a consideration of *Conservation in a Changing World*, Mace *et al.* (1998) indeed stress the importance of *preservation of process* rather than preservation of state – based very much on the viewpoint expressed by Smith *et al.* (1993). Considering the lessons from the past, Huntley (1998) concludes that population losses act to reduce a species' intraspecific genetic diversity, and thus the ability to respond to change. Reduction of habitat loss and persecution are thus to be seen not so much in terms of a loss of species and/or preferred states but a *loss of species' ability to respond to change* (see also Huston (1994), *Biological Diversity: The coexistence of species on changing landscapes*). Preservation of habitats is thus to be seen not so much in terms of arresting succession but in terms of capacity building. This perhaps brings us back to our initial consideration where the same thing goes under another name – rather than scrub clearance being seen in the light of favouring sets of species, such as the butterflies of Wicken Fen, It should be seen in terms *of maintaining habitat, and therefore species, diversity which can facilitate response to change*. A monoculture of scrub does not maintain this diversity. However, it is also clear that larger blocks of habitat may also be important, as discussed by Shafer (1990) in: *Nature Reserves: Island theory and conservation practice* (Figure 5.3). This is, though partly in relation to the fact that larger areas might be more heterogeneous, with more

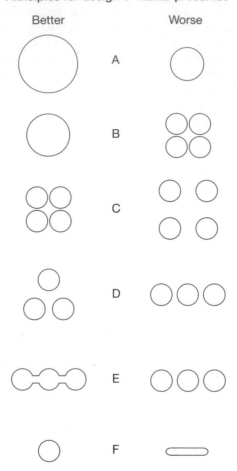

Figure 5.3 Nature reserve patterns (from Shafer, 1990). Larger areas are preferred to smaller and proximity and connectivity are seen as important.

transitions from, say woodland to glades, involved. Shafer's conclusions include the following:

1. The more land set aside, the more species are preserved; smaller reserves are liable to be more vulnerable to stresses.
2. Species numbers increase with size of area but tend to diminish beyond a certain point. Nature reserve size should be influenced by the territory size of the largest wide-ranging mammals – this then serves as an "umbrella" for other species.
3. Habitat fragmentation and nature reserve insularization should be avoided (but see item 8 below).

4. Boundaries should not create conditions which discourage animal movement.
5. Buffer zones between reserves and surrounding areas with human land uses are desirable.
6. If larger reserves are not possible, the smaller reserves should be inter-connected to encourage migration, though not all of the population needs to be able to migrate.
7. Reserves should be replicated to reduce vulnerability.
8. If natural habitat connectivity is not possible, corridors should be created.

To these ends: "*laissez-faire* management should be the exception rather than the rule".

A good example of item 2 above is that adopted by the Royal Society for the Protection of Birds (RSPB) in the UK for the bittern (Tyler *et al.*, 1998). Here, there are few birds in the UK and only at a few sites. The RSPB have radio-tagged and tracked individual birds to assess the size of their ranges at different times of the year. This research enables recommendations to be made for the appropriate sizes and locations of alternative sites which may be identified and managed to encourage the spread of this rare and vulnerable species.

In summary, intervention is necessary and this to facilitate not only diversity but also movement in times of changing conditions so that the populations have the "reservoir" capacity to withstand change and also the mobility capacity to respond to change.

These kinds of analyses seem sound but do seem to be nested in the idea that "surrounding" (often agricultural) land will be rather bleak, with few species – which is no more than the "island" approach embodies. A more fundamental approach is adopted by Huston (1994), who concludes that "the fact that agricultural productivity and biological diversity are strongly influenced by the same environmental conditions *should* provide an economic rationale for the preservation of biological diversity. The hopeful scenario is one where agriculture and conservation are not incompatible, and indeed where capacity building involves building the capacity of the countryside as a whole (see Lowe *et al.*, 1997) rather than in the small area of nature reserves.

What is also interesting is the existence of parallel thoughts in social science. Agarwal writes in his treatise of "green villages" of decentralization to allow for local self-organization just as Pahl-Wostl talks of decentralized management to allow for ecological self organization. Additionally Trudgill and Richards argue for local calibration of management policies. These two approaches are indeed not just in parallel but come together in a key book by Berkes and Folke (1998), *Linking Social and Ecological Systems: Management practices and social mechanisms for building resilience*. The authors write of how "environmental surprises" can arise but also how these may not necessarily be damaging if there is a social resilience which gives people a capacity to absorb such external changes. This might imply that the current state is a desirable one and that the system can be maintained in a particular state despite changes when, in fact, resilience might have a wider definition of having the ability to adapt to changed states which might inevitably occur. This is, however, a theme which is of crucial importance but of much wider significance than the context of nature conservation, so we will return to this in the concluding chapter.

5.4 Urban ecosystems and gardens

First, we must realize that it is an entrenched attitude that *cities are divorced from nature* (Plate 5.2): Thoreau wrote that "nature flourishes far from towns where people reside". In *The Environmental Revolution* Nicholson (1970) wrote of "modern man, corralled in narrow corridor streets" and of "the deformities imposed on man by exchanging a healthy natural open-air environment for such an unhealthy artificial indoor container as an apartment or an office". If we accept this, we may thus readily identify with the beleaguered naturalist in the city: Henry Williamson (1922), in his autobiographical *The Lone Swallows*, records the importance of his glimpsing wildlife in London – the romantic poet "enraptured by life's beauty" as he sees himself. He quotes another romantic nature writer, Richard Jeffries: "Where man goes nature ends". Williamson writes how he saw a barn owl in London "hunting in the very heart of turmoil" and then how "My fatigue passed, and hope came into my heart" and how he would be "as indifferent to his surroundings as my barn owl was to them". It is interesting that he refers to the owl as "mine" and how he sees the owl as emblematic of a lost nature in the city. Allied to this, is the importance of parks and gardens, not only as "reservoirs" of wildlife but also in terms of contact with nature.

Plate 5.2 Cities: New York, USA. It is an entrenched attitude that cities are divorced from nature. However, parks and gardens are important not only as "reservoirs" of wildlife but also in terms of contact with nature.

On the other hand, while many earlier nature writers have in the past dwelt within a rather bleak scenario of cities, the future looks more hopeful. For a start, Roberts (1966) has already written in *The Penguin History of Europe* that while the individual can be robbed of individuality in huge, anonymous cities, many people have actually had a greater freedom of choice through moving to cities. More recent developments act to stress the human scale, public participation and contact with nature. Davis (1998) talks of "greening the urban desert" in the *Ecology of Fear: Los Angeles and the imagination of disaster*, contrasting the "ugly city" with "wilderness beauty" but then seeing urban communities with structures, paving, grass, shrubs and trees as an expression of the reintegration of man and nature. *The current trend is one very much of resisting the construct of man separated from nature in cities but one of the importance of green space* "greenways" (green corridors) and of gardens and gardening as shown by Smith *et al.* (1988) in *Greening the Built Environment*. The authors see the importance of integrating ecology, environment, health, welfare and economy.

Second, we must also realize *cities are dependent on other ecosystems*. Go through your food store, if you can gain access to it, taking a map of the world and plot on the map the country of origin of the goods. For terrestrial products, the chances are that most continents will be represented and that most products will be from domesticated environments. But even if you were totally dependent on goods from your local area, you would still be dependent upon agricultural ecosystems discussed below. The point is that we are still intimately connected with and dependent upon agriculture, the countryside, nature and wilderness, however much we sit in air-conditioned, centrally-heated rooms. In cities, we do not have immediate contact with other ecosystems but through television, reading and transport of goods we are involved with ecosystems. The links are ones of consumption and of mental constructs of nature.

A further point is that we can also construct *cities themselves as ecosystems*, seeing the same kinds of energy, water and nutrient flows as well as biodiversity considerations as in any ecosystem. In 1980 Bornkamm *et al.* (1981) reported on an urban ecology symposium, discussing fluxes of organic carbon, flora, habitat diversity, invertebrates, soils, birds, snails, foxes, trees, grasses, lichens and hydrology, and these cover not just those enclaves of "nature" such as parks and gardens but also many other types of site as well. Woodlands are often to be found within cities presenting sites for wildlife, recreation and also for economic use (Grey and Deneke (1978) and also Jones and Talbot (1995) report on the potential for coppicing in Sheffield, UK, Figure 5.4).

Cities are places where human cultural activities dominate and virtually everything structural is the result of a decision but also where species colonize and take advantage of opportunities just as anywhere else. There are thus both deliberately created (see Nicholson-Lord (1987), *The Greening of the Cities*) and opportunistic elements of the flora and fauna of cities as described by Gilbert (1989) in *The Ecology of Urban Habitats* and Laurie (1979) in *Nature in Cities*, together with many opportunities for nature conservation (see Kendle and Forbes (1997) in *Urban Nature Conservation*, which describes appropriate practices and techniques).

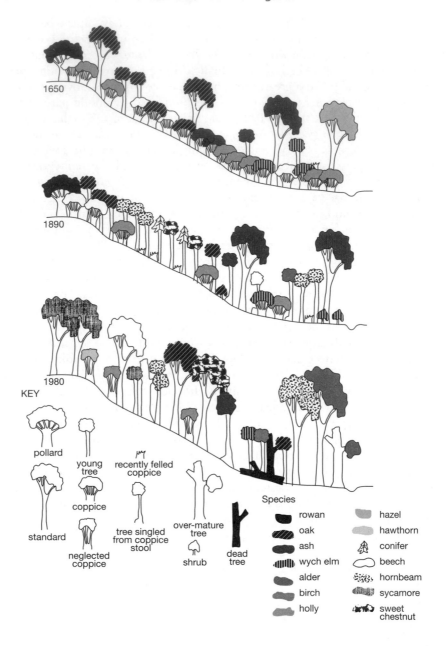

Figure 5.4 Coppicing in woodland in Sheffield (from Jones and Talbot, 1995). In 1650 the woodland is intensively coppiced and pollarded but this has decreased by 1890. By 1980 the woodland is no longer used for wood products. The trees are all old, of roughly the same age, and thus face an uncertain future without regeneration and there is little in the way of any shrub layer; in some ways protection and non-use can lead to decline. Why not use them as a source of wood products, valuing the woodland as a source of products as well as in terms of nature?

The special features of urban ecosystems were studied by Rebele (1994). The author follows Tansley (1935) in seeing ecosystems as "mental isolates" for "the purposes of study" and sees the city as an assemblage of various ecosystems at a number of scales at the level of organisms, populations and whole communities and dependent upon the area of "unsealed" space. A distinctive feature is that the proportion of successfully established species introduced (deliberately or not) from other biogeographical realms is higher in cities than in rural or forest areas, with greater immigration than extinction. This alone justifies an interest in urban ecology in pure biodiversity terms and it is interesting to note that some cities might be of more interest than some "wilder" areas which, while they might be more "natural" might have lower species numbers. Rebele sees this as a special feature of the high species richness of cities when compared to the countryside. I might, however, relate this to a comparison with some monoculture domesticated systems in agriculture or forestry as opposed to some less managed ecosystem. Rebele explains the richness in terms of the high habitat diversity of urban and industrial areas. This, in turn is related to man-induced disturbance which initiates the colonization of disturbed or newly-created habitats. *Human activity can thus increase biodiversity rather than, as is often assumed, decrease it.* This parallels the observation of Johns (1997) who in a study of timber production and biodiversity in forests wrote that "the assumption that logged forests support none of the biodiversity of unlogged forest is plain wrong" and indeed can cite cases where the diversity of the former is greater than the latter. The debate may, however, then become one of "native" versus "introduced" species and whether numerical biodiversity alone is desirable, as rehearsed by Perlman and Adelson (1997) when discussing values in conservation (see Section 5.3 above).

In discussing nature conservation, above, we concluded that for mobile species to be able to adapt to new, more appropriate, locations under conditions of change then the provision of the facility to move was important, especially with consideration of "corridors" of appropriate habitat. Corridors are also very much a subject of urban nature. This was stressed by Adams and Dove (1989) in *Wildlife Reserves and Corridors in the Urban Environment.* Such concepts have been adopted more recently as "Greenways". Here there is much more than just the idea of the facility for species to move but also is a consideration of the "fabric" of the city. Walmsley (1995), in an American study, writes how greenways can "stitch together" fragmented cities and urbanized areas with greenways, greenbelts and green spaces together making a comprehensive "green" infrastructure. Greenways are seen as being able to make powerful shapers of urban form at both macro- and microscales. Stormwater channels can often make the basis of greenways which link one area to another (McGuckin and Brown, 1995), improving ecological connectivity and porosity and, as these involve water, they use the term "blue–green" open space. Turner (1995) appeals for a diversity of greenways and in the context of London identifies several varieties of greenways including blueway, ecoway and cycleway.

The downside of green spaces in cities is that they can often be associated with the fear of crime: murders, rapes and muggings and other incidents often

taking place in parks, especially at night, away from well-lit areas where vegetation provides cover for a potential assailant. Nasar *et al.* (1993) conducted a
study on college campuses in relation to fear of concealment and blocked
prospects (limited views of what was ahead and limited prospects of escape).
The findings, involving fear-maps, showed that fear related physically to the
presence of trees, shrubs and walls, especially when near to pathways. Making
areas where paths cross parks more well lit, and at least open, could thus be a
means of keeping otherwise valued green spaces while reducing the element of
fear. Luymes and Tamminga (1995) felt, in a study of green spaces, that concerns for safety mean that many such areas are regarded as inaccessible. Opening
up areas and providing clear sight-lines may compromise ecological integrity.
However, careful planning of areas, providing visibility but also solitude without isolation was not incompatible with ecological values – areas could be both
green and safe. Thought has thus to be given to the layout of parks in cities and
the simple existence of a park is not universally or automatically a "good thing".
This is also shown by the study of Solecki and Welch (1995) who concluded
that green spaces could also act as "green walls". Parks located between socio-
economically distinct neighbourhoods can actually function as boundary landscapes. The park then impoverishes both neighbourhoods because when the
park acts as a boundary this often leads to less use of the open space resource
which can then become derelict. Clearly, the planning of green spaces should
include not only ecological considerations but also landscape architecture, layout and site positioning in relation to psychological and sociological factors.
Indeed, many would argue that these latter considerations should come first,
with ecology and species considerations coming second.

In terms of positive psychological factors, it is clear that we have deliberately
created many open spaces in cities and above all planted trees. Davies (1988)
writes of "the evocative symbolism of trees", a symbol having both historical
significance and physical features. Trees, the author writes, "have a creative
metaphysical capacity and a physical manifestation of ideas" and goes so far as
to say that there is "an elective affinity between trees and human thought". We
may thus talk of improvements to microclimate, purification of the air and
other mechanical effects of trees but it also seems instinctive to plant them even
when we substantially create an environment. Interestingly, in this context,
Porteous (1996, p. 137) in *Environmental Aesthetics* relates a study by Ulrich
(1984). Ulrich had earlier worked on natural scenes and their positive psycho-
physiological effects and in a subsequent control experiment it was found that
patients in a hospital recovered more quickly when they had a view of trees from
their window (the control view being a brick wall), implying that more positive
attitudes were engendered.

That gardens are therapeutic is not in doubt. The words "agreeable, shady,
cool and plants to give pleasure" are used by Curl (1997) and in a newspaper
gardening column (The *Independent* 19 June 1999) one gardener is reported as
talking of "aura, atmosphere, soul, spirit, solace and healing" when discussing
her garden. Porteous (1996, p. 134) reports the findings of surveys that 60% of
4000 gardeners cited feelings of peace and tranquillity as the most important

satisfaction to be had from gardens. For some low-income urban dwellers, the collectivity, self-esteem and sociability was also stressed. Diversity is also stressed in new urban projects, not just an open area in which to stroll, as stressed by Smales, (1999) when writing about mixed-use spaces. These should have children's play areas alongside woodland, wetland and other habitats, interspersed with paths and lakes. Mixed spaces, the author feels, "offer opportunities for creativity, e.g. community arts projects), involvement (allotments, gardening) and experience (sensory trails), rather than just a place to be". The essence of the planning is to involve the public and destroying "the myth that a productive landscape defies aesthetic sensibilities". In these contexts, Ghazi (1999) also reports on the benefits of community involvement in garden projects, giving people a renewed sense of self-esteem and worth, especially some groups of offenders. Here, their feelings of purpose are renewed. Some more objective measures of benefit, including lower blood pressures and less medication as well as more relaxation and the feeling of being needed are also reported by physicians in a senior citizen's centre with a garden project. Porteous writes: "Both urban gardening and the passive contemplation of vegetation appear to have a healing quality. Plants are long-term, they are steady in growth, they stay in place, they are quiet, their rhythms are slow and natural" and says that research confirms "long-held beliefs about the life-enhancing value of contact with nature".

There is, however, an element of struggle in gardening as well as of peace. The amount of effort involved in the struggle ranges from the "sometimes extravagant", as commented on by Wheeler (1998) and "Honey, it's three AM – What are you doing in the garden?" Editorial to "*The 20-minute gardener: the garden of your dreams without giving up your life, your job, or your sanity* (Christopher and Asher, 1996). There are also parallel notions of control and of "the wild". It is a common lament to be heard about gardens that they "can get out of control" yet at the same time we are ostensibly celebrating nature and naturalness. Malitz and Malitz (1998) have a book called *Reflecting Nature: Garden designs from wild landscapes*, which now contrasts with earlier versions of the ordered and formal garden. There thus seems to be a duality about gardens that is also present in the wilderness construct – wanting nature and wanting control. This, to me, does not present a paradox – *gardening is an expression of the heart of the man–nature relationship and we derive much pleasure of being involved with nature – but on our own terms.* A garden must be the ideal in the sense that we can have both involvement and control, whereas those who detract from the wilderness bemoan the loss of human control, which is precisely why those who celebrate the wilderness go there. What is also interesting is that we can invent and control gardens and thus what we produce tells us a lot about our nature, even if we chose not to control them closely. It is also interesting that we are back to the themes we raised when discussing nature reserves – ones of putatively preserving nature and also of expressing our preferences.

We might then be able to place ourselves along a spectrum of relationships with nature (Figure 5.5 along with expressions of preferences, Figure 5.6 and expectations, Figure 5.7). As we move from wilderness through a fruitful

Figure 5.5 Attitudes to nature. The ambivalence is that comfort and security are valued but we relinquish these as we move to the right of the spectrum. Great value is placed on wilderness but attitudes diverge – with those who espouse nature and become involved and physically committed to it, often seeing kind of merit in the loss of home comforts (e.g. backpackers) and those who espouse nature but don't get too involved if it means relinquishing comfort (e.g. enjoying a spectacular view from a car but not actually walking in the wilderness). Any one person is not necessarily committed to any placement on this spectrum consistently (e.g. the camping holiday and return to home). Gardens represent for many the acceptable face of nature where it is there but putatively and largely on our own terms.

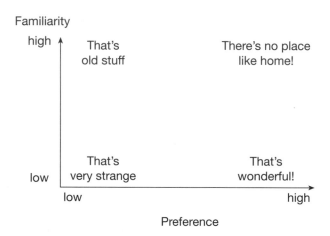

Figure 5.6 Familiarity/preference matrix from Porteous (1996). Familiarity can lead to boredom or security while unfamiliarity can bring fear or delight, depending on a person's psychological preference.

countryside to a tended garden to a formal flowerbed laid out in a park to a room in a house, we can move from fear of the uncontrolled-but-valued-for-the-natural through to a controlled, comfortable-but-maybe-boring-and-artificial spectrum. Gardens occupy for many people an acceptable balance between the two end-members – the wild and the controlled. More uncontrolled than the house but not too much so, in contact with nature, but not lost in it. Security comes from routine and recognition, which is why holidays and trips to the wilderness are exciting and stimulating, but we are usually glad to get home. Gardens offer both security and stimulation – and as we said when talking of wilderness, the importance of enjoying the sublime at limited risk to

Recreational activity clusters. Note the relatively small number who participate in active pursuits. (Drawn from data supplied by Parks Canada.)

Canada		Quebec	
People consuming no recreational type	19.1%	as for national survey	17.9%
People who only drive for pleasure	15.3%	as for national survey	18.4%
Picnicking and driving for pleasure	13.2%	as for national survey	13.6%
Sight-seeing, picnicking, visiting parks, historic sites, driving for pleasure	11.5%	swimming, driving for pleasure	10.3%
Swimming, picnicking, driving for pleasure	16.9%	swimming, picnicking, driving for pleasure, walking	10.3%
Swimming, picnicking, visiting parks, sight-seeing, driving for pleasure	7.4%		
Sailing, climbing, walking–hiking, tennis, picnicking, sight-seeing, swimming, driving for pleasure, visits to parks	13.2%	swimming, photography, driving for pleasure, picnicking, walking, visiting parks	13.9%
Participated in all activities mentioned	3.4%	as diversified activities	6.2%

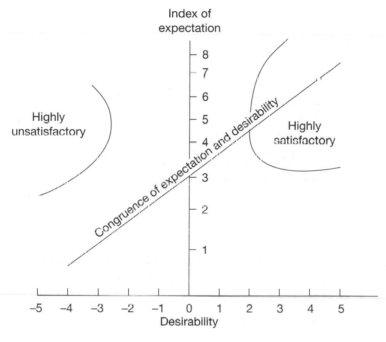

Figure 5.7 Preconception/desirability interactions, in this case in the context of the recreation experience (O'Riordan, 1976). The difficulties experienced by managers is the range of views of what is expected and what is desirable, ranging from taking pleasure in driving and/or picnicking/historic sites/walking/swimming and so on.

yourself is clear. Different motivations obviously then mean that different people may gravitate to either end members from the couch potato to the backwoods man while many in society range along the spectrum at different times, though with a tendency to live towards the controlled end. Real wilderness is perhaps something to be cherished but, for most people, only engaged with on visits. Porteous, indeed, reports that people felt better knowing that national parks existed even if they were non-users. Nature reserves are interesting in this spectrum because they are putatively wild but actually quite controlled. Wilderness is the negation of human achievement which some thus abhor and some thus relish. Nature reserves are the constructed value-laden manifestations of our respect for nature. Gardens are the acceptable face of nature.

Insights into the place of gardens in the human mind can be found in the useful compilation by Wheeler (1998) in *The Penguin Book of Garden Writing*. Here, fruits and vegetables are seen as important but more in terms of "efforts made by men and women of all ages to make beautiful the pieces of land they wake up to every day". It is interesting that a multiplicity of constructs are allowed from a "tamed" construct, like a clipped hedge or shrub to a "profusion", "riot of colour" or "tangle" and many gardens are physically constructed of hard structures and softer masses of plants, allowing both order and informality.

Of the wealth of many writings in *The Penguin Book of Garden Writing* I would pick out the extract (p. 21) from an article by Mirabel Osler (Osler, 1990):

> ... gardens are a paradox. They reflect their owners; they are totally dependent (on them); and yet in no time at all they are breathing with their own lungs, growing at their own pace, behaving with either wilful disregard or subjugation. Subjugated gardens abound, and I can see why. Unless discipline is maintained from the moment the spirit level is laid across the earth, you are nurturing a vast, tactile, heavy-scented siren which will keep you forever in its thrall. Happy and doting, you never stop planning for its well being ... to do without one is utter deprivation. You cut yourself off from a whole spectrum of sensory ferment ... discover ... with what brilliance and with what almost unpremeditated intuition you have created something of overwhelming beauty.

Clearly, the experience of beauty and of sharing in creation runs deep. The book also quotes (p. 26) Sir Francis Bacon "God Almightie first planted a garden. And indeed, it is the Purest of Humane pleasures. It is the Greatest Refreshment of the Spirits of Man; without which Buildings and Palaces are but Grosse Handy-works". One is also reminded of:

> The kiss of the sun for pardon
> the song of the birds for mirth.
> One is nearer God's Heart in a garden
> than anywhere else on earth.
>
> (From *God's Garden*, Dorothy Frances Gurney, 1858–1932)

Gardens may thus be seen as mystic and relating to nature in terms of creativity (like seed sowing) and are deep in our psyche. In a fascinating book (*The American Lawn*), Teyssot (1999) charts the cultural history and current usage of lawns. They can be seen as symbols of domesticity, leisure and wealth (not having to use all land to produce food to survive) and relate to accomplishment and to order. In the context of the early days of the colonization of America, there is also a discussion of the selection of species which will survive in different regions of America, a theme which is again relevant in terms of climatic change. That we are attached to gardens, and especially lawns – even in hot climates – is evidenced by Smith (1998) in *Water in Australia*. In Perth and Canberra respectively (Figure 5.8), 42% and 55% of domestic water use is out-of-doors with 38% and 52% being used for garden watering (the remaining 4% and 3% is used for pools). In hot summers, *"some 90 per cent of water use in Canberra is applied to lawns"*.

The cultural commitment to gardens and especially lawns is self-evident. Even in the face of adverse conditions, gardens are created and energies focused on them. They give a conditional contact with nature. Self-expression is allowed even though there can also be a sense of inferiority in relation to expert opinion. Gardens must be at the heart of the man–nature relationship and we are liable to maintain them, whatever environmental changes there are. Wilson (1992) writes in *The Culture of Nature* how a great variety of idiosyncratic ideas may be expressed in gardens and finds it very telling that nature is lavishly replicated in shopping malls and other new human urban developments: "As primitive landscapes have vanished from the planet, we've surrounded ourselves with replications of them". In the Cambridge University Botanic Gardens there is a symbol of attachment to gardens in the form of the "Dry Garden". A range of plants has been installed and the garden is never watered; the study is to see which

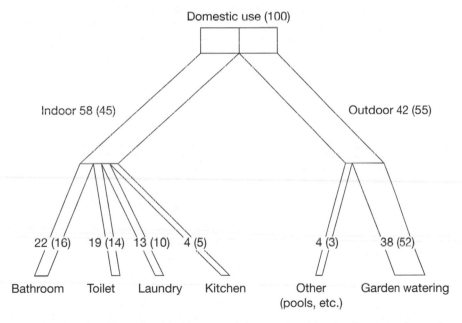

Figure 5.8 In a dry climate, it might be expected that people might abandon water-demanding lawns. However, cultural preference can over-ride this logic as the example of Australia shows. Figures are for average annual percentage use in Perth (with Canberra in brackets). It can be seen that about half the yearly domestic water use in Canberra is on watering gardens and in hot summers some 90% of domestic water in Canberra is used for watering lawns.

plants survive and flourish in a drier climate. This to me is a symbol of human tenacity and commitment to the garden.

Interestingly, Walton and Bridgewater (1996) writing *Of Gardens and Gardeners* extend the notion of gardening and justify a new perspective of the whole biosphere as a garden and people as gardeners, seeing the landscape as a biocultural region in which there are also protected areas. The wilderness then perhaps becomes a reverse metaphor for the "wild bit at the end of the garden"? The garden is thus also a metaphor for man–nature relationships, covering the range of enjoyment of not only control and achievement but also of a celebration of uncontrolled life forces and beauty. If this is true, then there seems little wrong with seeing the earth as a garden? Certainly, while the structural response of planners is to develop the green spaces in cities, we still seem to need the idea and reality of our gardens and contact with nature for therapy and escape from the pressures of life.

A WEEKEND WALK OUT OF THE CITY

Beset and full of meaninglessness,
fraught from expectations,
feeling hemmed and scheduled.

(Feeling like an artist painting for money
        than from the mind's eye vision,
or the poet completing for the deadline
        than for the thought and words).

(1) With inspiration boxed and bordered –
        the vocation trimmed to targets,
(2) insight rejected – the judgement must
        be others',
(3) the instinct swamped and sidelined –
        numerical criteria now account,
(4) creation curbed by drudge – but
        invention still expected.
(5) Ideas struggle out, but measured means
        restrict.

A tight cage has descended
and deep pits well,
stressed by stress
distracted by distraction.

(A songbird flutters crashed
below implacable glass
it can neither see nor understand.)

Distressed and hounded but
there is still some dim memory
which jerks and spurs towards the road.

The car takes the enclosing bubble shroud
to where others park.
Half remembered strangely familiar boots
that you as someone else used to wear
take your feet.

On the path out,
the enclosure still complete
head racing with hurt;
staring with unseeing eyes.
The mind turned inwards
with words and people,
rehearsing deep grooves of conversations
wherein the gridlock
of emotions with impossibilities
cuts a path deeply furrowed.
Trees pass distantly unreal
and views are shallow.

But then the steady rhythm footfall senses
        back
a song thrush pierces through
and a moss is noticed.
Now a stone has attention
then a flower or grass.
Focus and reality grow
while air is breathed.
Sitting on the grass, the strands are sensed
the trees have texture.

(Rain falling on a parched dry moor
at first beads and runs, rejected by the dry
        peat crust

then gradually the soil moistens and
accepts,
water infiltrating the softening soil
staining it darker, absorbing thirstily.)

Here and real.

(The spring comes through unwanted
choking stones
now muddy and silt-laden
then clears and sparkles.)

The clear sparkling lake finds an echo inside
and feeds inner meaning
into the husk of everyday.

(The shell discovered by the hermit crab.)

On the homeward path
seeing each milestone of hurt on the way
out
where there were thoughts of this and that
you smile and wonder why it mattered.

Expectations and money
lie now harmless with the waters of the
ditch
going nowhere.

Schedules, targets, criteria and deadlines
lie rotting in the rhododendron leaf mould
dark and temporary.

Insight, inspiration and instinct
affirm from the unmoved trees
shine from the unassailable flowers and
waving grasses
and sing with the therapeutic thrush.

The glass is gone
the wounded songbird only stunned
now flies, battered –

then light, relaxed and laughing free:
Going back to nature.

Further reading

+.+ Key reading
* Important reading
Recommended reading

Adams, W.M. (1996) *Future Nature: A vision for conservation.* Earthscan.

Adams, W.M. (1997) Rationalization and conservation: ecology and the management of nature in the United Kingdom. *Transactions of the Institute of British Geographers,* 22, 277–291.

* Braun, B. and Castree, N. (eds) (1998) *Remaking Reality: Nature at the millennium.* Routledge.

Bryson, B. (1997) *A Walk in the Woods.* Black Swan.

Cronon, W. (1995b) The trouble with Wilderness or getting back to the wrong nature. In Cronon, W. (ed.) (1995a) *Uncommon Ground: Toward reinventing nature.* W. Norton.

Johns, A.G. (1997) *Timber Production and Biodiversity Conservation in Tropical Rain Forests.* Cambridge University Press.

Krakauer, J. (1998) *Into the Wild.* Pan Books.

* Mace, G.M., Balmford, A. and Ginsburg, J.R. (1998) *Conservation in a Changing World.* Cambridge University Press.

** Perlman, D.L. and Adelson, G. (1997) *Biodiversity: Exploring values and priorities in conservation.* Blackwell.

*   Porteous, J.D. (1996) *Environmental Aesthetics: Ideas, politics and planning.* Routledge.

Proctor, J.D. and Pincetl, S. (1996) Nature and the reproduction of endangered space: the spotted owl in the Pacific Northwest and southern California. *Environment and Planning D: Society and Space.* 14, 683–708.

Part II

Soil: a fundamental resource

Chapter 6

Concepts of soil

Summary of key points:

1. Cultural constructs in the literature refer to soil in terms of "earth mother"; what is missing is the concept of soil as having an origin.
2. Soil is a basic resource which can be viewed as renewable, especially the organic components.
3. However, with regard to the mineral components, the rates of formation are very slow and much slower than many rates of erosion.
4. Soil is therefore best seen as a non renewable resource, deriving from the legacy of the past and which should be cherished as such.

6.1 The significance of soil

It's good to lay your head in the dead leaves
and smell the earth and sleep
under a bush in the breeze,
under the sky and sun

To lay down among gnarled roots
where the tree's bent sinews
cling to the rock, white and
white on grey,
to merge with the soil
and sleep like a twisted root yourself
seeking out the moist dark supportive earth.

The love of dirt is among the earliest of our passions, as it is the latest. Mud-pies gratify one of our first and best instincts. So long as we are dirty, we are pure … To own a bit of ground, to scratch it with a hoe, to plant seeds, and watch their renewal of life, – this is the commonest delight of the (human) race, the most satisfactory thing a man can do.

Charles Dudley Warner, 1876, from Wheeler (1996)
The Penguin Book of Garden Writing

Culturally the terms "soil", "land" and "earth" can be imbued with notions of yield and fertility: "mother nature", provision and abundance: "The earth shall endure and blossom forth in spring". Mahler, *Das Leid von der Erde* and with notions of attachment: ownership, patriotism and nationality. There are also negative notions to do with dirt – as illustrated by the fact that I was once asked to write an article under the umbrella title of "soil is not just dirt" (Trudgill (1978) "The earth will endure and blossom forth in spring." *MIMS Magazine*, an article on soils for the medical profession).

Writing about a group of Kenyan people, the Gikuyu, Mackenzie (1998, p. 24) quotes the words of Kenyatta written in 1938: as agriculturalists they depend entirely on the land, but as well as their economic life, land tenure is fundamental to their social, political and religious life. "It supplies them with the material needs of life through which spiritual and mental contentment is achieved. Communion with ancestral spirits is perpetuated through contact with the soil in which the ancestors of the tribe lie buried." They consider the earth as the "mother" of their tribe; "it is the soil that feeds the child through lifetime; and again after death it is the soil that nurses the spirits of the dead for eternity. Thus the earth is the most sacred thing above all that dwell in or on it". In a more modern society, this approach may now be somewhat overstated, but the notions of "mother earth" are still well established.

Other notable writings about the soil include the novel *The Growth of Soil* by Knut Hamsun (1935) described in the blurb as "the story of Isak, 'the tiller of the ground, body and soul', sinks its roots into man's deepest myths about his struggle to cultivate the land and make it fertile". It describes how he worked on the land, "there were stones and roots to be dug up and cleared away, and the meadow to be levelled ready for next year" and of the pleasure when the fields and meadows were looking well. Here the notions of struggle, mastery and productivity are embedded in the text.

In the 1880's novel *La Terre* (translated as *The Earth* (1980)), the writer Émile Zola has a central character very attached to the land – and land owner-ship forms a central theme in the novel. On revisiting his land, once lost and now regained, the text tells how:

> ...[he] stood for a long time contemplating [the field]: it was still there and seemed to be in good heart, nobody had harmed it. His heart overflowed with joy at the thought that it was his again, and forever. He stooped and picked up a lump of earth in both hands, crumbled it, sniffed it and let it trickle through his fingers. It was his own good earth, and he went home humming a tune, as though intoxicated by its smell.

These kinds of construct probably represents the closest attachment to the soil. At the other end of the spectrum, notions such as "British soil", "Russian soil" and so on now appear less frequently in the political arena and for many people remote from any direct involvement with the soil it is probably the notion associated with dirt which occurs to them.

Physically, soil science books construct soils as a fundamental resource upon which plant life depends, though the dependency is somewhat mutual as plant and animal remains contribute to the constitution of the soil.

Rachel Carson (1962) wrote in *Silent Spring*: "The thin layer of soil that forms a patchy covering over the continents controls our own existence and that of every other animal of the land. Without soil, land plants as we know them could not grow, and without plants no animals could survive." (See Plate 6.1.)

Plate 6.1 Soil profile. Rachel Carson (1962) wrote in *Silent Spring*: "The thin layer of soil that forms a patchy covering over the continents controls our own existence and that of every other animal of the land. Without soil, land plants as we know them could not grow, and without plants no animals could survive."

Green plants need sunlight for photosynthesis, together with carbon dioxide from the atmosphere, and also nutrients and water. While some nutrients and water can be absorbed direct from the atmosphere (especially nitrogen gas through fixation in root nodules), the soil is the principal store of nutrients and water for terrestrial plants. The mineral particles, often derived over long periods of time through the processes of erosion, deposition and weathering, provide the basic soil texture which, in turn, influences many fundamental properties especially soil structure and porosity. Weathering processes within the soil releases plant nutrients. The organic component is also a source of nutrients and an influence on soil structure, porosity and water holding capacity.

Our knowledge of soils developed principally in the nineteenth century with the development of agricultural chemistry and through pedalogical studies of the origins of soils through weathering (Yaalon and Berkowicz, 1997). One of the earliest notions was that of the Greek botanist and philosopher Theophrastus (327–287 BC) who wrote that soil is the source of plant nutrition. Vergil (70–19 BC), a Roman poet and farmer, made what is widely regarded as the first recorded observation in soil science that if you dig a pit and then attempt to replace the excavated soil, the soil which would not all easily fit back in would be the best (being heavier, i.e. clay-rich). In the seventeenth and eighteenth century the concepts of heavy and light (sandy) soils were still the dominant ones (Yaalon and Berkowicz, 1997). Knowledge soon grew from the mid-nineteenth century with notions of the responses of plants to the additions of chemical and organic fertilizers, the response of plants to soil types, the movement of soil water, the variation of soil types and their mapping and of the effectiveness of different management techniques.

With a more scientific agriculture, the movement of people to towns and the differentiation of labour, agriculture changes from being something that most people might be involved in to the specialism of the farmer, with specialist knowledge. Thus, a general spiritual attachment to the earth, such as recorded above with African people, declines but there is still a romantic attachment to the land felt by many people though one can differentiate between the attitudes and values of the non-farmers and the farmers directly involved.

The constructs of a spiritual attachment to the soil contrast markedly with constructs associated with the history of agriculture in the United States. Whitney (1994, Chapter 10) readily cites the "Dust Bowl" and the descriptive epithets of "earth butchery", "predatory agriculture", "spoilation" and "exhaustion" in what is seen as the "most rapid rate of wasteful land use in the history of the world". This approach to the land seems nested in the construct of North America as a wilderness to be conquered and of unlimited productivity as described in *Wilderness and the American Mind* (Nash, 1973), together with the assumption that there were always new fertile lands to the west. Indeed, it is instructive to look at the post-Dust Bowl era literature on soil conservation which still appealed to the pioneering spirit and the need for toughness of approach (see Inset 6.1).

The question arises of why such a contrast between the spiritual attitudes to soil resources and our use (i.e. abuse) of them should exist. Is it the dismissive, negative notion of "dirt" or is it the role of mechanization?

In the latter context, the answer may be one concerning the relationship between farmers and the erosion of the soil on which they depend. There is an interesting observation in Steinbeck's (1939) novel *The Grapes of Wrath* where a farmer climbs aboard a tractor and loses contact with the earth (perhaps mechanization has caused a change in attitude to vital resource?):

> He could not see the land as it was, he could not smell the land as it smelled; his feet did not stamp the clods or feel the warmth and power of the earth. He sat in an iron seat and stepped on iron pedals ... He did not know or own or trust or beseech the land ... He loved the land no more than the bank loved the land ... Behind the tractor rolled the shining disks, cutting the earth with blades – not ploughing but surgery, pushing the cut earth to the right ... Behind the harrows, the long seeders – twelve curbed iron penes erected in the foundry, orgasms set by gears, raping methodically, raping without passion ... And when that crop grew, and was harvested, no man had crumbled a hot clod in his fingers and let the earth sift past his finger tips ... The land bore under iron, and under iron gradually died; for it was not loved or hated, it had no prayers or curses.

It may also be that the "town mind" which Laurens van der Post (1986) spoke of in *A Walk with a White Bushman* when talking of Jung, is more at the heart of the matter:

> ... he was always close to nature ... he ... discovered that there was such a thing as a town mind, and that the whole of life was divided between town and country

Inset 6.1   Soil conservation propaganda from a US Department of Agriculture Soil Conservation Service 'Land of Washington' pamphlet (USDA, 1964)

There is a basic appeal to the pioneering spirit (a), constructing American toughness as appropriate for tackling soil erosion as it was in the 'frontier' days of the pioneers. Constructs of 'the good life' (b) and (c) relate to correct soil conservation measures (d), disaster follows those who ignore the message (e). Images of domesticity and correctness follow (f, g, h), ending with a sense of harmony and wonder (i).

(a)

(b)

(*continued*)

(*continued*)

(c)

(d)

(e)

(f)

(g)

(*continued*)

(*continued*)

(h)

(i)

minds … Western Europe was once governed by a country mind – even city states had a keen country awareness – but the country mind has very much diminished. People who know nothing about the land suddenly sit and rule like gods over the fate of other men. They think that they can pass a law and create a new state of being. Any farmer will tell them that you cannot do that, there are no short cuts; that everything is a process of growth, and growth has got its own time and laws and seasons, and is not man's time.

Farmers can still cause erosion so, alternatively, and given also that erosion and soil degradation can occur without mechanization, it may be that the soil is regarded as an infinite resource and bountiful and thus needing no special care? In this sense, it seems that soil may be grouped along with the other elements as air and water. Certainly, this is far from the case – soil is certainly only naturally renewable over very long time scales and is a finite body with definite origins and with a finite existence. To me, this is the central point. "Soil" and "Earth" are celebrated in literature and folklore as part of "mother nature", yielding and bountiful but thereby as an infinite resource – and something which can be taken for granted, and indeed dismissed as dirt. What is lacking is the notion of *origin* – there is no conception that soil came from somewhere, has an origin and is therefore actually finite and can be lost: the fact that there is no notion that soil has an origin is what is lacking in our cultural constructs of soil.

6.2 Soil renewability and sustainability

The idea that a resource can be viewed as renewable does usually depend upon the time scale over which it might be renewed. A geological resource, such as oil or coal, might be viewed as renewable but only if we wait for several millions of years of geological time for new deposits to form, so obviously from a human perspective these are regarded as non-renewable. On the other hand, a biological resource such as grassland can grow back within a few weeks of cutting and thus is regarded as a renewable resource. Soil resources occupy a peculiarly half-way position in this range of timing of renewability: their mineral components form over geological time scales of millions to thousands of years whereas their organic components evolve much more quickly.

The mineral components derive from weathering processes, either *in situ* or through the processes of weathering and erosion in one place and then by transport and deposition elsewhere. Although erosion and deposition can be very sudden, say during floods, most of the time many of these processes are very slow and gradual and in some areas were more active in the past than they are at the present day. In this sense, the mineral components can be regarded as non-renewable. However, the organic components, derived from dead and living plants and animals, are formed by processes which are often very active at the present day and in this sense they can be thought of as renewable. A few centimetres of organic soil can build up in a few years when pioneer vegetation, such as mosses, colonizes bare rock surfaces and earthworms can turn a layer of fallen autumn leaves into soil organic matter in the course of a few weeks. The organic components can also readily be replenished by the addition of manures. On the other hand, some organic components in soils have been dated as being several hundred years old and so, as with the mineral components, we might in part be trading on a legacy of the past.

Is soil, then, a renewable or non-renewable resource? This is a critical question since the answer will act to determine our attitude towards it as a resource we use. If it is renewable, then we need not worry so much about soil erosion and loss by other means because the soil will re-form; if it is non-renewable then we must carefully cherish our resources as once lost they will not be renewed again. The discussion above indicates that the answer is not initially a straight-forward one. In some respects the soil can be renewed: the rapid deposition of mineral matter in floods, the build up or organic matter and so on. In other respects it might be regarded as non-renewable if it is the product of weather-ing over thousands of years.

My answer to the question is, however, straightforward in that I would argue that despite the evidence for some renewability, it is foolish to regard soils as a renewable resource. This is simply because if you study the soil resources of the world which we now depend upon, most of them – both in terms of their areal extent and depth – are the legacy of past processes. Only a small proportion could be said to be actively forming at the present day. It is true that they are actively being modified at the present day but most relate to thousands or hundreds of years of weathering and deposition. The present resources thus relate to prolonged past weathering and climatic changes with their associated glaciations and downslope movements. Even the organic matter may relate to past processes in that proportions of the organic matter in some soils may be several years and sometimes hundreds of years old. If these portions are lost, they may not be readily replaced without considerable application of fresh organic matter which may then have different characteristics (for example, more readily decomposible and therefore lost more easily) when compared with the older organic matter. It follows that if we lose these resources, they cannot be easily recreated.

There is a fallacy that there is something called "an acceptable rate of soil erosion" when losses are matched by the rate of soil formation. During soil formation, larger particles of rock fragments and their primary minerals break down into secondary minerals such as clays and oxides to form a finer-grained soil. However, current rates of mineral soil formation *in situ* may be only 1–2 mm a year through such weathering and they are often a lot less (Table 6.1). Many soil erosion rates are greater than these rates of formation, often in association with a loss of vegetation and through cultivation.

Even if the rates of weathering were to keep pace with rates of erosion, weathering only produces mineral soil. Erosion is often of the most valuable organic rich and fertile top soil. While the organic component can be renewed

Table 6.1 Rates of weathering for selected minerals. Time taken for a 1mm crystal to dissolve (from Goudie (1998) based on the work of Lasaga)

Mineral	Years
Quartz	34 000 000
Muscovite	2 700 000
K-feldspar	520 000
Other feldspars	6 000–80 000

more readily than the mineral component by further additions and inputs, this itself represents unsustainability as the inputs have to be created from elsewhere. A sustainable system can be defined as one which does not depend on externally generated inputs but has a self-renewing capacity at a particular site. For example, a farm that recycles all the organic matter it produces back onto the land without any extra inputs can be seen as sustainable but one which is dependent on the import of (expensive) fertilizer and organic matter from elsewhere can sustain production but this does not represent sustainable use of that particular farm site.

The key issue is therefore not so much one of renewal or non-renewal nor of the acceptability of rates of erosion, but more one of sustainability: In most situations, erosion represents a loss of both organic and inorganic components which, although often recoverable, can only be replaced, respectively, from external sources or very slowly. Erosion is thus seen as an unsustainable net export of fertility from a site and a progressive loss of the soil resource.

This is especially so because the soil resource is largely a result of a sequence of past processes. Therefore, in order to understand the sustainability of the soil resource in more detail, we should study and understand the past sequences of soil formation: these sequences are unrepeatable in the foreseeable future and have given us our unique legacy of soils.

6.3 Soil erosion and soil formation

Central to the issue of whether or not soil erosion is significant is the notion that soil erosion can be balanced by soil formation. Indeed, where such a balance is achieved, this has been termed the "acceptable" rate of erosion. This notion should be examined for two reasons. First, the question should be asked as to whether, and if so under what conditions, such a balance might be achieved. Second, there is the point that soil erosion usually removes the topsoil, with its accumulated organic matter and associated fertility while any weathering which leads to the re-formation of soil will only provide inorganic material. It is true that the processes involved will lead to the release of inorganic nutrients like calcium, magnesium and potassium which are valuable to plant nutrition, but vital nutrients, especially nitrogen and to a considerable extent phosphorus, are not so released as they have minimal presence or availability in inorganic matter. These are held largely in forms available to plants in organic matter. If the surface soil containing organic matter is lost, the source of these vital nutrients is lost and is not replaced by weathering.

Many of these effects can be seen as detrimental, though not necessarily so. O'Riordan (1976) has made the distinction between:

- Exploitation – where there are short-term benefits but often an unsustainable long-term detriment.
- Conservation – where there may be fewer short-term benefits but greater long-term sustainable benefits (Figure 6.1).

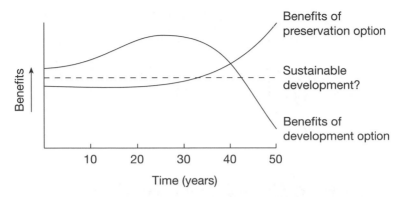

Figure 6.1 Development options appear attractive because they give better immediate returns. The hypothesis of this graph is that options which preserve resources, like soils, lead to greater eventual benefits as development inevitably lead to resource degradation (O'Riordan, 1976). It might now be argued that development and preservation are not mutually exclusive and a horizontal line might now be drawn starting on the vertical axis between the other two lines and continued horizontally, labelled 'sustainable development. (Modified from O'Riordan, 1976.)

De Boodt and Gabriels (1980) show that yields are closely related to soil depth (Figure 6.2). Soil depth decreases as erosion occurs meaning that at first the situation might be recoverable but eventually this is not so (Figure 6.3). While it is true that nutrients can be replaced by the addition of fertilizers, and so the effect of erosion can be recoverable, this is at the cost of the addition of fertilizers. Fertility can thus only be sustained at a cost and not by natural processes.

6.4 The legacy of the past

Many soil resources are a legacy of past processes. It follows that if we lose these resources, they cannot be easily recreated. Current rates of soil formation may be only 1–2 mm a year through weathering. Many soil erosion rates are greater than this, often in association with a loss of vegetation and cultivation. Erosion rates greater than these slow rates of formation lead to an unsustainable situation and a progressive loss of the soil resource. The present resources relate to prolonged past weathering and climatic changes with their associated glaciations and downslope movements. We should study and understand the past sequences of soil formation because these sequences are unrepeatable in the foreseeable future and have given us our unique legacy of soils.

The most significant origin of many soils in temperate latitudes were the Pleistocene glaciations: great amounts of pre-existing weathered material were removed from high latitudes and altitudes to be deposited at lower latitudes and altitudes. It is these mechanically weathered and transported deposits which formed the parent material of many soil resources. In lower latitudes, it is the prolonged duration of chemical weathering, often over many millions of years,

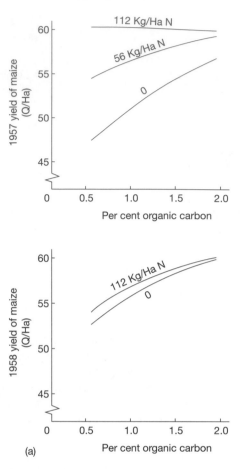

Figure 6.2 (a) Reasons maintaining the organic content of soils (i.e. adding manures instead of or as well as inorganic fertilizers). With no fertilizer, yield increases strongly with a rise in organic matter, adding fertilizer can reach enhanced levels of yield but one might note in the upper diagram that 2.0% organic carbon reaches nearly the same yield as a high dose of fertilizer but at a fraction of the cost. Soil depth and yield. In the lower example, expensive fertilizer makes little difference.

which has given us our soil resources. From these simple statements, it is already evident that without further glaciations or prolonged weathering we are using soil resources which cannot easily be re-created.

Plate 6.2, a picture of a glaciated landscape (Mount Cook in New Zealand) is really a picture of soil in the making. Morainic material is deposited at the base and snout of the glacier (Plate 6.3) and the material is largely unsorted in terms of particle size. Ground moraine, at the base of the glacier, can be very variable in size but tends to incorporate more finer clay particles, while ablation till, formed during melting of the surface of the glacier, tends to be of coarser,

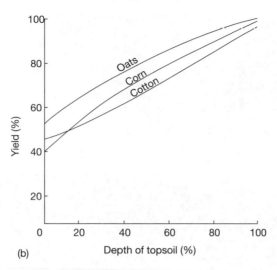

(b)

Figure 6.2 (*continued*) (b) Reasons for preventing soil erosion. Soil depth shows a high positive correlation with the productivity of many crops for reasons of greater moisture and nutrient availability. Soil erosion decreases soil depth. (From de Boodt and Gabriels, 1980.)

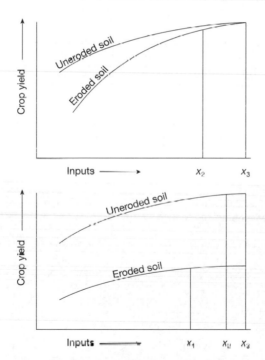

Figure 6.3 Crop yield and soil erosion. "Recoverable" soil erosion. Given that yield increases with increasing (fertilizer) inputs, it is evident that eroded soil can achieve the same level per input but only at a high level of (expensive) inputs. In the irreversible situation, however high the inputs, yields can never reach those of an uneroded soil, indeed a high level of (expensive) inputs hardly matches the yield level of the uneroded soil with minimal inputs. (From de Boodt and Gabriels, 1980.)

Plate 6.2 Mount Cook, New Zealand – soil in the making through glaciation.

often sandy, particles. Any meltwater tends to sort particles, the coarser deposits not having moved so far as the finer particles which tend to be deposited at greater distances, for example in lake basins.

Plate 6.3 Iceland: moraines left by glaciers form soil from material transported by glaciers from the mountains.

One further process which is significant in areas adjacent to glaciers is the deposition of fine-grained material (loess). This is derived from areas with scant or no vegetation (Plate 6.4) and may be deposited in deep deposits (Plate 6.5) or detectable in rather thinner layers within soils.

Plate 6.4 Mount Cook, New Zealand. In the early stages of deglaciation, lack of vegetation means that large amounts of fine wind-borne material can be transported by the wind.

Plate 6.5 In parts of Belgium, south of the last Quaternary glaciation, there is evidence of the accumulations from the processes shown in Plate 6.4 with large depths of wind-blown material termed loess.

Once the ice sheet has retreated, frost action dominates geomorphological processes (Plate 6.6) leading also to movement of soils on steeper slopes.
Where such downslope movement is very active, any vegetation which becomes
established is easily disturbed, with the disruption of any surface mat of
vegetation.

Plate 6.6 Iceland: in the early stages of deglaciation, frost action splits up larger boulders into
smaller grain-size material.

With an amelioration of climate, freeze–thaw action declines and the soil
material assumes a more stable position. Vegetation can now become more
easily established (Plate 6.7 and Plate 6.8). Two things can now occur. First, the
important organic processes which sustain the soil system can occur, especially
the recycling of minerals through root uptake and their return to the soil
through leaf litter decay; organic matter can also accumulate and be incorporated into the soil. Second, once the soil material is no longer being disturbed
mechanically, the movement of rainfall through the soil facilitates the dissolution of easily soluble material in the topsoil and its movement to the lower
layers of the soil. The soil is now set to develop and differentiate according to

Plate 6.7 Iceland: vegetation then becomes established, though can still be disrupted by frost-heave.

Plate 6.8 Iceland: with climatic amelioration, stable mats of vegetation can form.

the topographical and climatic conditions in which it is situated, as discussed in the next chapter. For this chapter, the notions of origin and effective non-renewability are seen as fundamental.

Further reading

```
    *    Important reading
         Recommended reading
```

Brady, N.C. (1995) *The Nature and Properties of Soils.* Macmillan.

Courtney, F.M. and Trudgill, S.T. (1984) *The Soil. An introduction to soil study.* Edward Arnold.

Curtis, L.F., Courtney, F.M. and Trudgill, S.T. (1976) *Soils in the British Isles.* Longman.

*   Houghton, J. (1996b) *The Sustainable Use of Soil,* Royal Commission of Environmental Pollution, Report 19, Cm 3165, HMSO.

Chapter 7

Soils and environmental change

Summary of key points:

1. Soil is differentiated according to its origin and because of subsequent developments involving interactions with the environment.
2. The processes operating are reflected in the arrangements of soil materials in layers or horizons.
3. It is those factors which relate to interactions with the environment which are self-evidently sensitive to environmental change, especially the hydrological, chemical and biological components.
4. These factors can also be readily altered by human manipulation, together with any losses of soil particles through erosion.
5. Discussions of global warming should not detract from the concerns we already have about human manipulations of soils and soil erosion.

7.1 Soil differentiation

Soils differ because of their inherited qualities relating to their origin (Chapter 6) and because of subsequent developments involving interactions with their environment. It can already be appreciated that soils will differ in grain size according to the origin of any deposits which make up their parent material. The next most important factor in subsequent development is topography. Soils can be much thinner on steeper slopes – as a result of past periglacial processes continued downslope moment; this inevitably leads to deeper soils on slope foot sites where deposition occurs. Drainage is also closely related to topography, wetter soils occurring not only on slope foot sites and low-lying areas but also on plateaux if the rock is impermeable. Given combinations of particle size distribution and slope position, all given in relation to past processes, the critical factors then becomes climatic, especially temperature and rainfall which influence the growth of vegetation (and hence organic inputs to the soil) and the rates of soil processes.

Soil sections tend to be visibly organized in horizontal layers, or horizons. The nature, presence, absence and distinctness of soil horizons provide strong clues to the operation of past and current processes.

Organic surface horizons may be thought to be related to high rates of organic matter input but in fact where they are highly visible they are more often related to low rates of output, i.e. low decomposition rates. Decomposition is inhibited where biological activity is low, usually because it is either too cold or too wet or both and/or too acid. A thick layer of accumulated organic matter immediately indicates that conditions are not conducive to biological activity. A thin layer of such material where there is significant input of organic matter indicates a high rate of biological activity and the incorporation of organic matter into the soil (Figure 7.1). Thicker organic layers are usually termed "O" horizons and mixed mineral – organic horizons are termed "A" horizons. Above the "A" horizon may be thin organic layers: "L" where the leaf litter is still recognizable; "H" for humus where the plant remains are not (and there is a possible intergrade between the two referred to as "F").

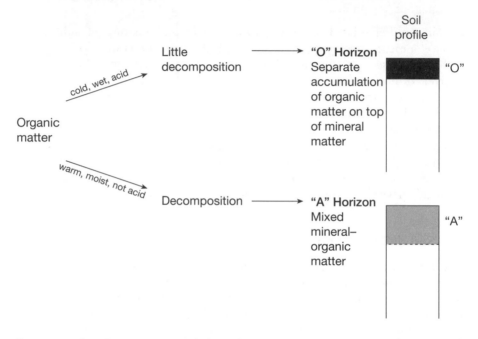

Figure 7.1 Organic matter accumulation: if biological activity is reduced (cold or wet), decomposition is less.

The brown earth is the most undifferentiated soil in that it has no marked horizons (Plate 7.1). Here, organic matter is incorporated into the surface "A" horizon and the subsurface or "B" horizon is not markedly differentiated. Other soils show more distinctive horizons, especially in terms of colour. If you ask the question "why are many soils brown or reddish in colour?" the answer is not only due to the incorporation of dark-coloured organic matter but mainly a matter of the presence of oxidized iron (or simply "rust"!). Oxidized iron, or Iron III (ferric iron) is red coloured whereas in waterlogged, anoxic soils the iron transforms to reduced iron, Iron II (ferrous iron) which is bluey-grey in colour.

Iron, however, is not only reduced under anoxic conditions, it is also reduced to Iron II under acid conditions (Figure 7.2). Such acidity may originate in the

Plate 7.1 Brown Earth.

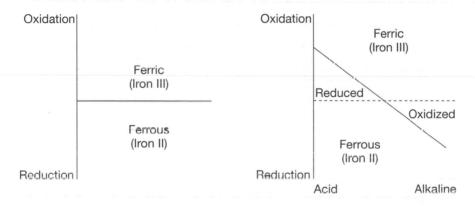

Figure 7.2 Oxidation and reduction. Reduction is exhibited at low pH as well as low oxygen.

organic layers of a podzol soil. Here organic acids are produced which reduce the iron. Iron II is more soluble than Iron III and hence can be mobilized into solution and thus down the soil profile in percolating waters. The iron may also become incorporated into the molecular structure of the organic acids (chelation). Further down the soil profile, as the soil becomes less acid and/or the organic acids break down, the iron may become redeposited as Iron III giving a red layer and thus the characteristic sequence of black–grey (iron reduced to iron II and washed out) – redder (iron washed in and deposited as Iron III) (Plate 7.2 and Figure 7.3).

Soil profiles, then, indicate a tremendous amount about soil processes, as discussed further in Courtney and Trudgill (1984).

Plate 7.2 Podzol.

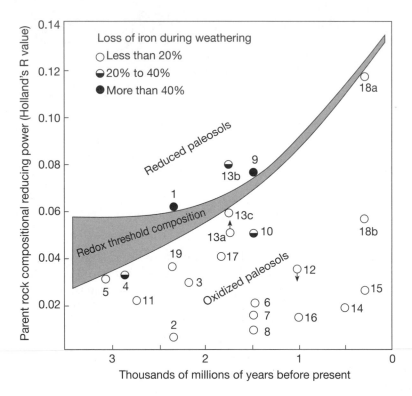

Figure 7.3 Oxidation occurred progressively during geological time as the earth's atmosphere evolved (Retallack, 1992).

7.2 Soils properties and processes

Soils are fundamental to plant growth and therefore to the living organisms of the planet, including, through crop growth and animal husbandry, to people. Soil fertility involves not only the nutrients which are present in the soil but also all the factors which influence plant growth, including rooting depth, water retention, soil particle size distribution (texture) and organic matter content. If these are impaired in any way through soil erosion (a loss of the actual soil resource itself) or a lowering of soil quality (through soil pollution, excessive water retention or drought or loss of nutrients) the capacity for plant growth is impaired and an environmental issue is perceived. The environmental issues thus perceived include soil erosion and other degradations of the soil resource, the effect of acid rain on soils, soil contamination and the addition of fertilizers and pesticides – including both their effects on the soil and the wider environment.

The key soil properties which are involved in altering the value of the soil resource and the wider impacts on the environment include soil texture, structure, porosity, water, aeration, organisms and nutrients.

Soil texture

Soil texture is a fundamental property from which many other properties can be predicted. In general terms, it involves the way the soil feels when handled where terms such as sticky and gritty or "heavy" and "light" may be used. More precisely, it relates to the particle size distribution, that is the proportions of sand, silt and clay (Table 7.1). Gritty or light soils have a high proportion of sand while sticky or heavy soils have a high proportion of clay. It follows that the former are easy to work (dig or plough), but not very retentive of water while the latter are difficult to work but are more retentive of water. Loam soils are made of mixtures of sand, silt and clay and benefit from having a range of properties which are useful to agriculture.

There is a good relationship between the feel, or "workability" of the soil and its particle size distribution. The precise soil texture can be analyzed quantitatively and plotted on a triangular diagram (Figure 7.4) identifying the soil in terms of the dominant particle size or the mixture. The diagram is asymmetric in that clay fractions tend to dominate the soil behaviour, even when present in relatively small quantities. During soil particle size analysis, the

Table 7.1 Sand, silt and clay particle sizes

	mm
Coarse sand	2–0.2
Fine sand	0.2–0.02
Silt	0.02–0.002
Clay	<0.002

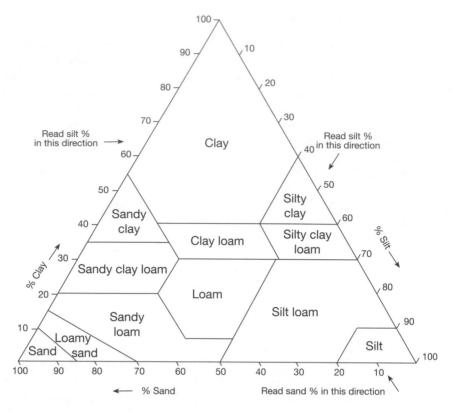

Figure 7.4 The texture triangle.

soil can be dispersed and then passed through a sieve to assess the relative importance of the coarser particles, but the smaller sized silt and clay are best measured using sedimentation (see Inset 7.1). Texture can be measured by the behaviour of a moist sample soil – sand is not coherent whereas clays can be rolled into a thin thread.

As Vergil showed (quoted in Chapter 6, from Yaalon and Berkowicz, 1997) there was an early appreciation of the relationships between soil particle size distribution and other soil properties. It is true that a heavy clay soil can be fertile, but also that it is difficult to work. In order to understand how soil particle size distribution influences soil fertility, it is important to understand how the mineralogy of particles may vary with size and also the properties of the different minerals varies. The first point to make is that soil texture refers to size alone and not to mineralogy, but the second is that each size fraction is usually dominated by particular minerals (Figure 7.5).

Sand-sized particles frequently comprise the most resistant minerals, especially quartz. However, they may be made of any mineral derived from a wide variety of rocks, for example, in a limestone area, sand-sized particles of limestone can be found. Commonly, however, the sand-sized quartz particles

Inset 7.1 Sedimentation

Soil material is stirred in water and after 5 minutes the larger sand particles are settled out; smaller silt particles take 8 hours to settle. A sample may be withdrawn from the remaining suspension with a pipette at the given times and the dry matter weighed, alternatively the density of the liquid may be measured using a hydrometer, and the proportions of sand and silt measured, clay be measured by subtraction.

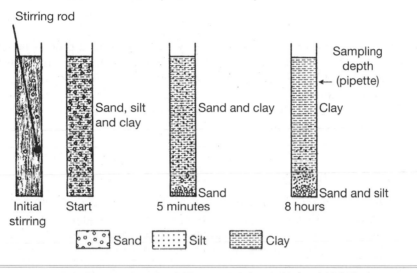

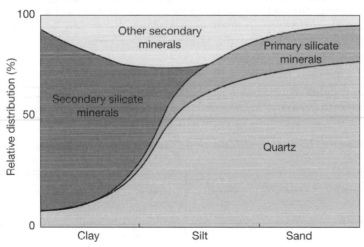

Figure 7.5 Particle size and mineralogy (from Brady, 1995). The terms clay, silt and sand refer to particle size, not necessarily to mineralogies, though each fraction is dominated by particular mineralogies: secondary silicate minerals are largely the clay minerals, silt is dominated by small particles of quartz and sand by larger particles of quartz but can also be sand-sized particles of any other rock or mineral.

impart their characteristics of being chemically inert, yielding little or nothing in the way of nutrients during very slow weathering. Silt-sized particles are often dominated by smaller particles of quartz and have similar properties. Clays, however are of different mineral types which may shrink and swell with the addition or subtraction of water and which have the property of retaining nutrients. Many nutrients which are present in the soil are in the form of positively charged cations (for example the cation calcium, Ca^{2+}, which can be derived from lime or calcium carbonate, $CaCO_3$, the compound of calcium carbonate being balanced between positively charged calcium and negatively charged carbonate). Clays have a net negative surface charge (see Inset 7.2 for explanation). It becomes apparent that the nutrient cations will become attracted to, and held onto, the negatively charged clay surfaces. Such a process is termed *adsorption* (as opposed to absorption, which is the term, for example, for the way water is held *within* a sponge, *ad*sorption refers to the *ad*herence of the cation *onto* the negatively charged surface).

The significance of this is that the adsorption is of such strength that the nutrient cation is held onto the clay and is not easily washed off by water

---

Inset 7.2 Charges on clay surfaces

Clays are minerals formed of layers of aluminium and silicon atoms arranged within a tetrahedron (or shape like a pyramid) of oxygen atoms (see Figure 7.6 for a diagrammatic representation). The silicon atoms at the centre of the tetrahedron become replaced by the more mobile aluminium atoms over time during weathering. Since silicon is Si^{4+} and aluminium is Al^{3+} there is an excess unsatisfied negative charge of one when this replacement occurs. This charge can be satisfied by cations absorbed between the layers of the clay mineral or by cations adsorbed onto the clay surface.

In some clays there is simply an "edge" effect – inside the mineral the charge balance is satisfied but at the edge there are "spare" unsatisfied charges (see Brady, 1995).

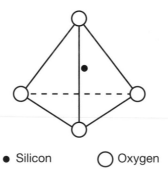

● Silicon   ○ Oxygen

Figure 7.6 Silicon tetrahedra (from Brady, 1995).

moving through the soil. At the same time it is held sufficiently lightly for plants to take up the nutrient from the clay. The presence of clays in a soil is therefore important to the nutrient status of a soil. Clay types vary considerably in their nutrient holding capacity, with montmorillonite, a more complex clay, having much greater nutrient holding capacity than the simpler kaolinite. In addition, organic matter also exhibits similar nutrient holding properties and the capacity is often much greater than that of the clays (Figure 7.7).

Thus, a knowledge of soil properties can be used effectively to predict the outcome of management actions. It can be predicted that positively charged cations will be held in the soil and not readily leached to runoff waters, as is the case with calcium (Ca^{2+}) and potassium (K^+). Nitrate is in anionic form (NO_3^-) and is therefore not held onto clay surfaces which are also negatively charged – hence the propensity of nitrate to leach into runoff waters. Similarly, sandy soils, without the negatively charged clay surfaces, will have very limited nutrient retention capacities, even when the nutrient is cationic. However, even clay soils may not retain nutrients because of the arrangement of their particles, that is to say soil structure is also important.

Soil structure, porosity, water and air

Soil particles are characteristically grouped into coherent aggregates termed structures or *peds*. These may be very weakly expressed, as in a loose, sandy soil, or strongly formed, as in a cloddy, clay soil. This already indicates that texture

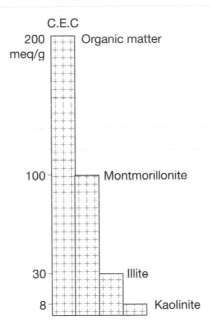

Figure 7.7 The relative cation exchange capacity of organic matter and three types of clay (from Courtney and Trudgill, 1984).

can have a strong influence on structure, the smaller clay particles having a tendency to stick together when wet. Clay soils also have a tendency to shrink and crack into solid lumps when dry. Other factors act to bind particles together – organic matter, organic gums, fungal threads (mycelia) and micro-organisms are all important; inorganic cements like iron oxides and calcium carbonate can also be involved.

The environmental significance of soil structures is twofold. Firstly, larger aggregates are more difficult to move by the forces of wind and water, hence a well-structured soil is less liable to be eroded. Secondly, the aggregates leave pore spaces between them through which water (and also air) can move. Thus, the link with cation retention discussed above is that while clay soils may have a high surface area for cation adsorption, a clay soil which has dried and cracked may become so porous by virtue of the existence of cracks as to present very little opportunity for adsorption – water simply runs down the cracks towards the base of the soil and then laterally to streams. Thus any nutrients derived from the surface of the soil (say from fertilizers) may run down the cracks with the water. In this way, the total pore space is not as important as the connectivity between the pores and it is the soil structures which are important in that they give rise to interconnected pore spaces between the soil aggregates.

Water is held very tightly in small pore spaces, but can move freely through larger pores. Surface tension and absorption are strong forces which dominate in the smaller pores but with larger pores the size of the water body increases and the pull of gravity on this volume becomes greater than the surface tension and absorption effects. Thus, water only moves freely through the larger pores.

Hygroscopic water is held in micropores less than 30 μm in diameter and may only be removed by heating and evaporation. Capillary water is held in medium-sized pores 30–75 μm diameter and can move by surface tension effects and thus by capillary action. In macropores , > 75 μm diameter, water is weakly held in the soil and can move under the influence of gravity, all the more so if the pores are connected to each other.

This latter point applies equally to air though gases can diffuse through the smaller pores as well. Gases may dissolve in water but movement is maximized when the pores are not water-filled. Oxygen is used up in soils through root respiration and through decomposition, the oxygen combining with organic carbon (C) to produce carbon dioxide (CO_2). Carbon dioxide can thus build up in soils, especially those of low permeability to air, but also escapes to the atmosphere where the soil is permeable.

The composition of soil air is thus lower in oxygen than the atmosphere but higher in carbon dioxide:

	Oxygen, O_2%	Carbon dioxide, CO_2%
Atmosphere	20.9	0.03
Soil air (common ranges)	15–20	0.25–4.5

In terms of greenhouse gases (Bouwman, 1990), soil produces carbon dioxide, CO_2, dinitrogen oxide, N_2O, ammonia, NH_3, hydrogen sulphide, H_2S and methane, CH_4. The first is produced as a result of respiration and decomposition, the second two as part of the nitrogen cycle and the latter two as a result of incomplete decomposition of organic matter under anaerobic conditions, especially in wetter soils.

Organisms

Some estimates of the total numbers of the larger soil fauna (including earthworms, insects and nematodes) have been made of around 60 million per hectare, a total weight of around 700–800 kg per hectare (Russell, 1961). When the amounts of micro-organisms are also considered, with bacteria at 1000 to 4000 million per gram (Russell, 1961), plus fungi, algae and other organisms, it is small wonder that people have been able to say that in a field where cows are grazing, there may well be far more live biomass invisible under the soil surface than there is of cow biomass visible in the field. While these estimates of figures have their uncertainties and can vary widely according to conditions, they do stress the concept of a "living soil".

In essence, these organisms function as an organic matter decomposition factory, releasing the energy fixed during photosynthesis and breaking down plant residues to their constituent carbon dioxide and water and also decomposing plant proteins to release nitrogen (Figure 7.8). However, other

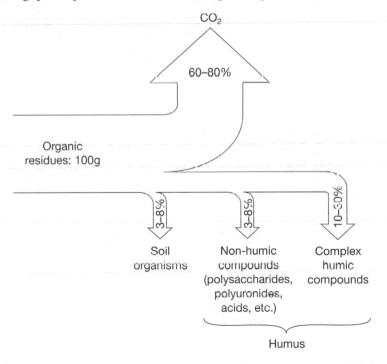

Figure 7.8 The fate of organic matter on decomposition (adapted from Ellis and Mellor, 1995).

substances are released, including other gases, together with by-products, especially of incomplete decomposition and also some substances can be created, especially nitrate through nitrogen fixation. Organism activity can also greatly enhance the release of nutrients through the weathering of minerals, mainly through cation exchange processes and the action of organic acids produced during partial decomposition processes. Indeed, many minerals are far more soluble under the action of organic acids than they are in water (Figure 7.9). Many apparently chemical processes, such as the acidification of water by carbon dioxide (Figure 7.10) are, in fact, biologically motivated as the carbon dioxide in soils is largely produced by organism respiration and the active decomposition of organic matter. In addition, organisms have a significant physical role to play in terms of binding soil particles together in soil structures and also in creating connected pore spaces through the action of burrowing.

Soil organisms are thus a key element in the terrestrial biosphere and sensitive to environmental change, especially if such changes lead to changes in soil moisture and temperature.

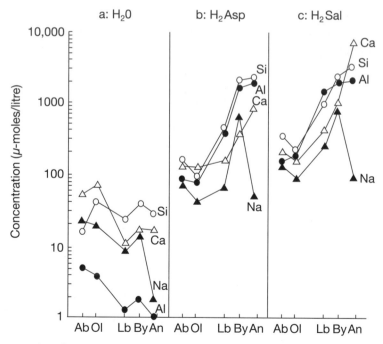

Comparative dissolution of cations from a series of minerals in (a) deionized water, (b) aspartic acid, (c) salicylic acid. Values for room temperature. Ab: albite, Ol: oligoclase, Lb: labradorite, By: bytownite, An: anorthite (from Huang and Kiang, 1972).

Figure 7.9 Most minerals are more soluble in organic acids than in water.

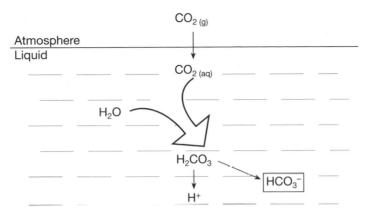

Figure 7.10 CO_2 and the acidification of water. An acid is defined as a substance which dissociates (splits up) in water to yield hydrogen ions (H^+) – it can thus be seen that carbon dioxide produces H^+, thereby acidifying water.

Soil nutrients

Many chemical elements are used as nutrients by plants, especially potassium, calcium and magnesium, mostly derived from the weathering of minerals, often organically controlled, but also from atmospheric particulate or dissolved sources. This is also true of the micro-nutrients required only in very small quantities as a trace element such as iron, copper, molybdenum, manganese and zinc. The key nutrients are, however, nitrogen and phosphorus. The latter can be derived from rock weathering but the former is derived largely by nitrogen-fixing bacteria either in root nodules or from free-living organisms in the soil, much of the sources of these two nutrients is derived by recycling through the decomposition of decaying organic matter. Biological processes are thus of key importance in the nutrient status of a soil. Nutrient status can thus be changed not only by the addition of nutrients in fertilizers but as much by changes in factors which affect biological cycling and other biological processes in the soil

Hydrochemical processes

In agricultural and semi-natural catchments, the quality of the stream water is a primary influence on the flora and fauna which is able to live in the streams and also a dominant factor in the usefulness of the stream water for human consumption. The waters may have characteristics which reflect the natural conditions of, say, soil and geology and also which reflect the influences of human activity, such as pollutant loadings, land use management practices and climate (Wilby, 1995). Streams which are laden with sediments, and very high nutrient concentrations or are of high acidity tend to provide poor conditions for wildlife and generally require further treatment for drinking purposes. Streams which are clear, of low to moderate nutrient status and neutral pH provide the best conditions for wildlife and the better drinking waters. Central to these considerations is the role of soil processes. For undesirable substances, such as acids derived from atmospheric deposition, the soil can act as a kind of "mediator" or "filter" between the external source and the stream, soil processes helping to minimize the impacts upon stream water. Alternatively, the soil can actually act as a source for sediments and substances in solution in streams. The critical factors are the hydrological pathways which exist through the soil and the chemical reactions which can occur along these pathways. The relationships between soil hydrochemical processes and climate can be assessed not so much by looking at the relationships with rainfall *per se* but at the relationships with rainfall intensity – a factor seen as changing under many climatic change scenarios.

Connectivity

The first important factor is the directness of the connection between incoming rain water and the stream. Simply, if water can run rapidly over the soil surface or even through the soil via cracks and large pores, it can carry any surface material with it directly to the stream. This water will thus take any atmospherically derived pollution, soil surface sediment or any surface-applied manures, sewage sludge, fertilizers or pesticides straight to the stream. Alternatively, if water soaks into the soil and flows gradually downslope through the soil, many surface-derived substances can be retained in the soil and minimize stream contamination. The other side of the coin is, however, that for any substance already in the soil the reverse will be true: surface or any other rapid flow will not pick up material from the body of the soil whereas slow percolation through the soil can eventually emerge at the stream bank high in concentration of soil-derived substances. This is illustrated in Table 7.2.

Table 7.2 The relationships between flow regime, contamination source and concentrations in streams (based on Trudgill and Coles, 1985)

	Rapid flow (surface or through soil macropores)	Slow percolation (through the soil matrix)
Surface substances (sediment, atmospheric pollutants, manures, sludges fertilisers*, pesticides*, humus acids)	*Transferred direct to stream* (**high** concentrations in stream)	*Retained in soil* (**low** concentrations in stream)
Soil derived substance (fertilizers,† pesticides,† soil mineral weathering products	*Retained in soil* (**low** concentrations in stream)	*Transferred to stream in soil throughflow water* (**high** concentrations in stream)

* directly after application while they are still on the surface and before movement into the soil.
† some time after application, after movement into soil.

It is clear that the division between surface-derived and soil-derived substances is not straightforward in that some substances (starred), like fertilizers, can initially be surface-derived (*) and later soil-derived (†). What is crucial here is the intensity of the rainfall. Heavy rain means both that (a) rapid flow is more likely than slow percolation and (b) that a surface-derived substance is washed off before it can become incorporated in the soil; light rainfall obviously means that (a) slow percolation is more likely and (b) that surface-derived substances can be washed gently into the soil and retained therein as a soil-derived substance.

Two factors thus emerge as important:

1. the timing and intensity of rainfall subsequent to the timing of any surface application; and
2. the intensity of the rainfall in relation to the rate at which water can be accepted into the soil.

Under item 1, light rainfall after a surface application is liable to be beneficial as it will encourage incorporation into the soil where it has more chance of being retained; heavy rains will increase losses to the stream.

Under item 2, we should study the relationships between rainfall intensity and the rates of water acceptance by the soil; both these can be studied in terms of relative rates: of rainfall intensity (in mm hr^{-1}) and the soil infiltration rate (mm hr^{-1}). The soil infiltration rate is influenced by four main factors:

1. **Soil texture**, for example sandy soils can generally accept water more rapidly than heavier clay.
2. **Soil organic matter** at the surface and vegetation type, for example an open, porous leaf litter under woodland can accept water faster than a denser humus mat under some other vegetation like heather moorland.
3. **Soil structure**, for example a soil which is well aggregated and contains numerous large pores or cracks can accept water faster than a weakly structured soil.
4. **Soil water**, for example a soil which is already nearly saturated with water will accept less water than a soil which is drier.

None of these factors operate simply or in isolation. Soils which have similar textures can have very different structures and organic matter contents in relation to vegetation types and land use management; clay soils when dry tend to crack and have much higher infiltration rates than when they are wet and not cracked; structures are not always stable and aggregation may change during the wetting of a storm; some dry soils can often initially repel water if they are baked hard or if there are crusts of algae or networks of fungal mycelia present which can show hydrophobic properties. Soil infiltration rates can thus be a result of complex interactions between a number of factors.

It is, however, possible to make a number of generalizations as shown in Table 7.3. Table 7.3(a) shows some characteristic values of infiltration in relation to soil texture (all other factors being equal) and Table 7.3(b) suggests

Table 7.3 Soil infiltration and rainfall intensities

(a) Soil infiltration and soil texture

Soil texture	Infiltration rate (mm hr^{-1})
Loamy sand	25.0–50
Loam	12.5–25
Salt loam	7.5–15
Clay loam	2.5–5

(Adapted from Gregory and Walling, 1973)

(b) Rainfall intensities

mm hr^{-1}	return period (years)*
10	<1 (i.e. frequent)
12	1
15	2
18	5
24	10
28	20
35	50
40	100

* Probability of occurrence based on records of average past frequencies, not an
actual prediction. Comparing (a) and (b) shows that the infiltration capacity
of many clay soils will be exceeded many times within one year, for other soil
textures, their infiltration capacity will only be exceeded by the occurrence of
more unusual storm events. Based on data presented by Briggs and Smithson,
1989 p. 193. Some global warming scenarios predict higher rainfall intensity
and thus less infiltration.

some characteristic values for rainfall intensities. Comparing the data in the
two halves of the table suggests that sandy soils can absorb most of the rainfall
intensities occurring, while silt soils will be more liable to produce infiltration
excess overland flow and clay soils will be less liable to accept commonly
occurring intensities. The purpose of this table is not to show that there is a
rigid relationship between soil infiltration, texture and rainfall but to illustrate
the likelihood of the overlapping values. Table 7.3(a) is somewhat misleading as
overland flow can occur on compacted sandy soils and rapid flow can occur on
cracked clay soils. Indeed, structure and porosity are often the key factors. This
is especially the case as macropores can greatly influence infiltration rates, giving
high values, even when the rate of the main mass of the soil is low (Figure 7.11).
Indeed, it is possible to predict from a general knowledge of infiltration and
percolation rates that a surface-applied substance may be safely retained within
the soil when, in fact, it may pass very quickly to the stream. This, indeed, has
been one of the major implications of research into soil water movement in the
past 20 or so years.

Matrix, cracks and macropores

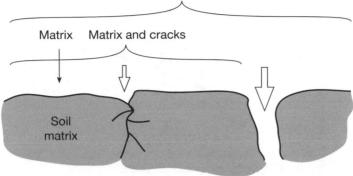

Figure 7.11 Infiltration of water into the soil. Due to the presence of macropores, the rate may be high even when that of the mass of soil is low.

Chemical processes

If water has the opportunity to contact the soil material, the substances may be lost from or taken up by the water. Substances already in solution in the water may be adsorbed onto exchange sites and/or react with material already in the soil; substances may also move into solution from the soil.

Surface-applied substances

- Acid deposition may be neutralized – acidification, buffer capacity
- Sulphates
- Nitrates soluble – not retained but can be taken up or transformed
- Organic matter substances decomposed
- Heavy metals not mobile
- Phosphate adsorbed, removed + sediment
- Pesticides – water soluble, absorbed, broken down depends on the breakdown rate

---

Soil-derived substances

- Nitrates
- Humus acids – uptake
- Soluble weathering products
- Plus all of the surface applied solutes once incorporated in the soil

---

The conclusions can be that there are critical combinations of soil structure and rainfall intensity which strongly influence the movement of solutes through the soil (see Trudgill and Coles, 1985). In general, at higher rainfall intensities surface-applied solutes are more liable to appear in soil drainage waters while soil-derived solutes are less likely to thus appear. Low rainfall intensity clearly

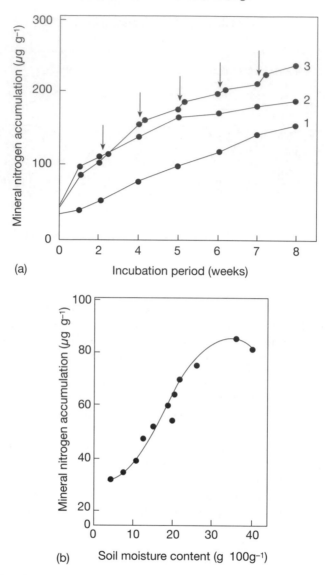

(a)

(b)

Figure 7.12 Combinations of soil moisture and temperature influence soil nitrogen transformations (from Haynes, 1986). (a): (1) fresh soil, (2) dried and re-wetted soil, (3) dried and rewetting in 5 cycles (arrowed).

favours the absorption of surface-applied solutes into the body of the soil. If rainfall patterns then change to ones of greater intensity then there are implications both for soil structural management and also for the timing of surface applications relative to rainfall events – and this is together with the probabilities of associated rainfall intensities, an increase in convective storms leading to more events of higher rainfall intensities. Value judgements then enter in on the desirability of the soil to stream transfer for each surface applied substance.

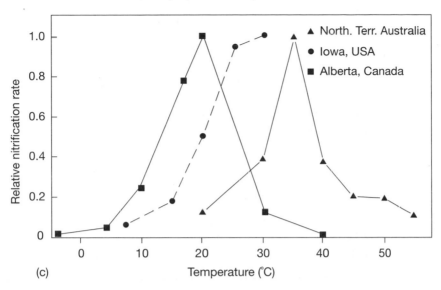

Figure 7.12 (continued)

In terms of the soil biological processes (see, for example, Fitter (1985) *Ecological Interactions in Soils*), combinations of soil moisture and temperature which favour life processes are the important factor (Figure 7.12). It is clear that there is an optimum combination of "warm, moist" with dry and waterlogged being sub-optimal, whatever the temperature and cold and hot being sub-optimal whatever the moisture content. The literature on soil organic matter and soil biology (e.g. Kononova, 1966; Haynes, 1986) provides ample illustration of these relationships. Generally, given adequate moisture, soil warming tends to increase rates of nutrient release (e.g. Luekewille and Wright, 1997) together with higher rates of organic matter decomposition, though enhanced root growth, and consequent increased nutrient uptake, does not necessarily mean that the released nutrients are leached out of the soil.

A good review is provided by Kirschbaum (1995) who reports on temperature dependence of soil organic processes and finds a considerable range of results. One generalization would appear to be that the sensitivity of soil organic matter decomposition is greatest at lower temperatures, with far greater increases per degrees of rise between 0 °C and 10 °C than at higher temperatures. The calculation is that at annual mean temperatures of 5 °C then a 1 °C rise would result in 10% loss of soil organic carbon, whereas the same rise at 30 °C would only lead to a loss of 3%. The percentage amounts also mask the fact that the actual quantities released at lower temperatures would be greater as cooler soils tend to have greater stores of organic matter due to slower inherent decomposition rates. However, Berg *et al.* (1995) also point out that the organic matter store in the soil represents the balance between litter input and decomposition rates. With enhanced growth in relation to increased carbon dioxide, there could be enhanced litter fall leading to increased storage if the conditions for decomposition are not optimal. These conditions should be seen not just in terms of soil moisture and temperature but also of whether the litter has a low inherent rate of decomposibility, often in relation to lignin content (Johansson *et al.*, 1995), and in terms of the C:nutrient ratio (McKane *et al.*, 1995).

In these ways, experimentation and extrapolation of known relationships can be made, generally indicating enhanced decomposition of soil organic matter but the combinations of factors are important and these can give rise to considerable variation in the overall balances which would result. We may now ask the question how sensitive soil properties and processes are to climatic change in order to assess the possible outcomes.

7.3 Sensitivity of soil properties and processes to environmental conditions

The scale of the expression of a feature depends upon:

1. the duration of the process forming it;
2. the susceptibility of the material being acted upon.

This is expressed in Figure 7.13. Here, large-scale features can only be produced by the long duration of process, but the time scale shortens if the material is more easily reorganized. Thus granite, weathering very slowly, will take many thousands or millions of years to show features adjusted to a particular set of

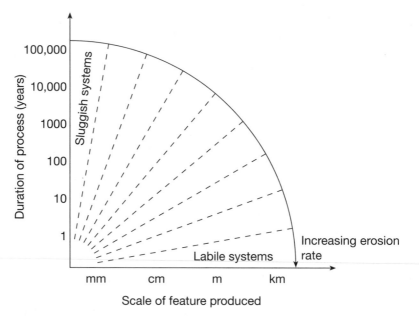

Figure 7.13 Scale of effect and duration of process.

conditions while the sand on a beach will adjust its form to the conditions of each tide lasting only a few hours. An important consideration is, however, that conditions are very rarely constant for thousands or millions of years and so large scale features characteristically tend to exhibit inheritance from previous conditions – as is the case with the UK upland glaciated landscape which owes most of its form to the last million or so years of glaciations than to the last 10 000 years of temperate conditions. These latter have not existed for long enough for the landforms to have adjusted to them and still display a glacial inheritance. There thus exists a lag in the system which the larger scale forms tend to display (Figure 7.14).

Lagged response

Change

Response

Figure 7.14 Lags and adjustments.

Similar considerations apply to soils. Figure 7.15 suggests that soil "A" horizons form in 10^2 years while deeply weathered oxic soils can take 10^5 to 10^6 years to form (see Inset 7.3 for explanation of soil terms). Thus the organic components of soil will be susceptible to much shorter-term fluctuations in

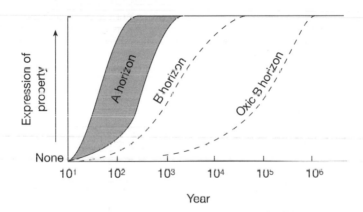

Figure 7.15 Time taken to produce soil features (from Birkeland, 1984).

Inset 7.3 Soil terms used by Birkeland (1984) in Figure 7.15

A horizon – mixed mineral and organic topsoil. B horizon – weathered and altered subsurface horizon. Oxic B horizon – characterized by a virtual absence of weatherable minerals and the presence of residual quartz sands and hydrated oxides of iron and aluminium.

conditions and the soil properties and components discussed above, the sensitivity to changes in environmental conditions will tend to decrease down this list: this is shown in Table 7.4 and Table 7.5 for sensitivities to environmental change and human manipulations.

In terms of the timescales over which nutrient dynamics and vegetation responses change in relation to temperature increases, some useful experimentation has been performed by Ineson *et al.* (1998a) (see Figure 7.16). They took cores from an upland site at 845 m altitude in the UK Pennines with a mean annual temperature of 3.5 °C and placed replicates at lower altitudes: 600, 480 and 171 m with respective mean annual temperatures of 5.1, 6.3 and 8.1 °C. In the first year spring-time losses of NO_3–N from the transposed soils ranged from around 0.25 to 1.0 mg l^{-1} for a brown earth, 0.5–1.25 for a podzolic soil and 1–4 for a peaty gley. In the second year losses from each of the soils were all close to a range of 0.25–0.5 mg l^{-1}. Although the rainfall decreased with altitude (from1605 mm per annum at the higher site to 1129 mm at the lower) it was concluded that this was not related to differences in water balance (and indeed the pattern was similar where an enhanced irrigation regime was used at a lower altitude site). Increases in denitrification were not implicated in causing lower levels of nitrate leaching, instead, the inference was that enhanced

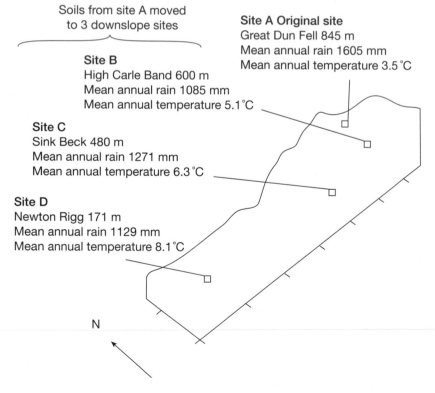

Figure 7.16 Transposing soil cores from upland to lowland areas (from Ineson *et al.*, 1998a,b).

Table 7.4 Senstivity of soil processes and properties to environmental change

Soil water and air	Highly sensitive to changes. Responsive on a daily basis, with rainfall, seasonally, yearly water balance and longer time, depending on site drainage
Soil organisms and their activities	Sensitive to changes: some micro-organisms activity daily with temperature and moisture, certainly seasonally and also progressively with changing conditions
Soil nutrients	Less sensitive, some response to temperature and moisture with the water but more seasonally in relation to water balance and biological activity plus progressively as net conditions change
Soil structure and porosity	Tends to evolve over time scales of a few to tens of years but may be more rapid if biological activity (and organic inputs) change more rapidly
Soil horizons and soil types	Mineral horizons generally evolve over hundreds to thousands of years and may not be adjusted to current dominant conditions but a net result of past processes, but organic horizons can react more quickly – generally a result of long-term net balances rather than short-term fluctuations
Soil texture	Almost a "fixed" property but does evolve slowly over a scale of hundreds to thousands of years, especially in more soluble material. May well not be adjusted to current dominant conditions but a product of past processes

Table 7.5 Senstivity of soil processes and properties to human manipulation

Soil water and air	Can be immediately affected by cultivation including greater water retention and less aeration through compaction and the reverse through ploughing and both also progressively in the longer term
Soil organisms	Can be stimulated through fertilizer addition or reduced by cultivation producing less suitable conditions; progressive reduction if organic matter is not added
Soil nutrients	Can be immediately increased by fertilizer addition; progressively lowering if crops removed and nutrients not replaced
Soil structure and porosity	As for soil air and water – structures immediately made smaller by cultivation, more plate-like by compaction and porosity similarly increased or decreased
Soil horizons and soil types	Can be homogenized down to plough depth, even with only a few ploughings, inverting upper horizons, deeper ploughing can break up sub-soil
Soil texture	Difficult to alter but clay soils can be better structured and more workable by the addition of organic matter over time and sandy soils made more retentive of moisture and nutrients also by organic matter addition. Cultivation can lead to erosional loss

vegetation growth in response to increased temperature at lower altitude resulted in increased nutrient uptake. The results showed that increases in mean annual temperature ranging from 1.6 to 4.6 °C resulted in a vegetational and hydro-chemical response within one year; indeed some decreases in leaching loss were apparent in the first few months of the experiment. Laboratory simulations using soil warming cables supported this, showing responses within five months (Ineson et al., 1998b). The conclusion was not only that increased temperatures (where adequate moisture exists) leads to enhanced vegetation growth and nutrient uptake with consequent decreasing leaching losses but also that the processes would be closely correlated in time with any changes, evidently responding without significant time lags.

The response of the organic fraction to the combined effects of temperature and moisture may be judged from Kononova (1966) where increases in tem-perature evidently lead to greater decomposition and carbon dioxide production provided moisture is adequate, the latter becoming a limiting factor as tem-peratures increase and moisture is reduced. Temperature rise may thus give rise to increased decomposition in wetter areas but not necessarily in drier areas, though the temperatures used to demonstrate this latter effect are higher than is likely to occur so this is not necessarily a realistic predictive statement. Thornley and Cannell (1997) present relevant modelling data on the effect of an increase in soil temperature of 5 °C in grassland (Figure 7.17). Total carbon is depleted following increased soil biological activity; as organic matter is decomposed, nitrogen is released from the decomposition of organic matter. It is predicted that the effect lasts for some 30 years with a 5–10% loss of soil total carbon.

At the other end of the time spectrum, dated sequences of palaeosols exist in the geological column which can give insight into the time periods over which various textural and chemical changes can occur (Retallack, 1990). Equilibrium between climates and soils may not exist, as very few soil features vary linearly with time and climatic conditions may not be stable enough for some features to necessarily develop. In addition, there may be some intrinsic thresholds which occur over time (Muhs, 1984). These include prior conditions which have to develop before a subsequent change in soil development can occur. These can be important in soil changes in the absence of environmental change but may also be involved during soil responses to environmental change.

In support of the non-linearity of soil development, Rockwell et al. (1985) suggest that it decreases exponentially over time (Figure 7.18) as the active or dynamic factors (fluxes of matter and energy, organisms, fluctuations in moisture status) act upon the passive factors (mainly parent material). These authors recognize that there is unlikely to be a smooth curve under climatic changes but a curve that changes to differing end points as climate changes, never reaching a steady state. They also include data on estimated ages of differing soil types, ranging from 8000 to 200 000 years – though this indicates the ages of the soils rather than necessarily the time taken to develop or respond to changing conditions.

However, it is possible to suggest minimum times for some soil features to develop from the data provided by Retallack. Not only is it suggested that soils

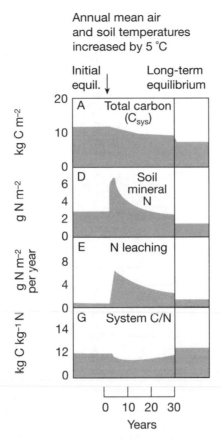

Figure 7.17 Results from using the Hurley Pasture Model on the effect of an increase in soil temperature of 5 °C in grassland. Total carbon is depleted following increased soil biological activity, organic matter is decomposed and nitrogen is released. The effect lasts for some 30 years with a 5–10% loss of soil total carbon. (Modified from Thornley and Cannell, 1997.)

appear at different times in the geological column according to an evolving geochemical environment and especially the appearance and development of plant life, examples from dated sequences suggest the timescales over which various features form. On alluvial terraces clayey horizons have been found to form in 40 000–100 000 years, the variation being in relation to a number of factors including climate (drier areas having greater inputs of windborne material leading to more rapid formation) (Retallack, 1990, p. 269). It could thus be suggested that textural alterations are thus possible over geological eras lasting millions of years (see Table 1.1) but where climatic fluctuations last over smaller timescales of less than 10 000–100 000 years, changes are unlikely to be reflected in an "equilibrium" textural response. In terms of weathering, data from Table 6.1 can be repeated here, with a range of time taken for a 1 mm crystal to dissolve (from Goudie 1998) of from 520 000 to 34 000 000 years

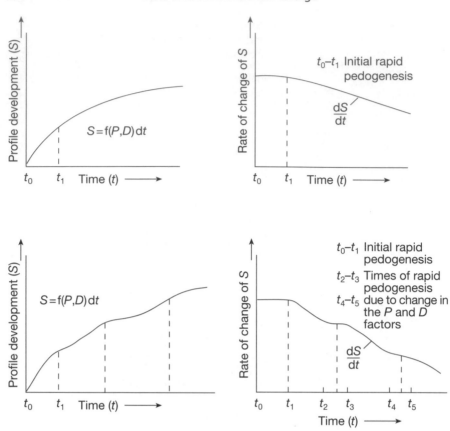

Figure 7.18 Non linearity in systems (from Rockwell *et al.*, 1985). S = soil formation, P = passive factor, D = dynamic factor.

showing that the same conclusion may be made for the minerals. Retallack also shows that weathering rinds in andesite and basalt of 1–2 mm thickness may take over 250×10^3 years to form. For other chemical changes, carbonate accretions range from 0.09 to 0.51, with a mean of 0.26 g cm^2 a^{-1} in a cool desert region in new Mexico

Intermediate between these rapid vegetation and hydrochemical responses and the slower textural and chemical changes are the rates of organic matter accumulation in peat soils. Data from Retallak is as follows, though these are liable to be maximum rates because compaction will occur as the peat thickens (Table 7.6)

We may generally conclude that vegetation and hydrochemical responses can be of the order of around one year, organic accumulations of about 10 cm of the order of 100–1000 years to form, chemical response times of the order of 1000–10 000 years (though with a wider range than this, depending on solubility and conditions) and textural responses of the order of 100 000–1 000 000 years (though again with a wider range, depending on conditions).

Table 7.6 Rates of peat accumulation

Peat thickness	Time to form (years)	Calculated yearly accumulation rate
0–4 cm	0–80	0–0.05 cm a^{-1}
4–40 cm	40–800	0.1–0.05 cm a^{-1}
40 cm–4 m	400–8 000	0.1 cm a^{-1}–0.05 m a^{-1}
4–40 m	4 000–80 000	0.1–0.5 m a^{-1}
40 m	>40 000	>0.5 m a^{-1}

Clearly, we may now redraw Figure 7.13 with some more realistic estimates of timescales of responses (Figure 7.19).

In terms of the sensitivity to human manipulation through agricultural activity, the timescales of response can all be shortened except for soil texture, though in that case the effects can be mitigated.

Agricultural conversion of forests and grasslands generally leads to an increased oxidation of soil organic carbon, with mean losses of soil carbon ranging between estimated means of 21–46% (and extreme values of 1.7–69.2%) of the total soil organic carbon (Bouwman, 1990, p. 68). Land use change can produce globally amounts of carbon dioxide estimated to lie between the ranges of 0.1–5.4 Gt C a^{-1} with a modal estimate of around 0.5 to 1 (Bouwman, 1990, p. 73).

We have, then, a picture of soil responsiveness to environmental change which covers an immense range of timescales. Factors related to soil water, organic matter and organisms together with plant growth can show responses within the range of days to a few years. At the other end of the spectrum factors relating to particle size and weathering cover a more geological timescale of thousands to even millions of years. Intermediate responses may be shown in terms of factors like soil horizons and soil structure. Human influences can

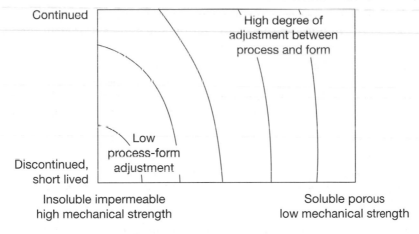

Figure 7.19 Figure 7.13 redrawn with time.

generally shorten the timescales of response including a wide range of agri-cultural and other land management practices as well as the exposure of soil to erosive agents.

In terms of meaning and significance, the more responsive properties can be most readily changed by human activity, can be the most readily altered – to both desirable and undesirable states – and also can be most readily renewed by altering and enhancing what Rockwell *et al.* (1985) termed the dynamic factors (fluxes of matter and energy, organisms, fluctuations in moisture status). The less rapidly responding properties can also be affected by human activity but cannot be regarded as easily renewable. We may alter the *effects* of soil texture and parent material without actually affecting the material itself and change the situation to a more desirable state, for example by improving the nutrient and moisture retention of sandy soils by adding organic matter, ameliorating clay soils by the addition of coarser and/or organic material and by draining them. Here there is no irretrievable effect and indeed the desirable state has to be maintained by constant renewal of the manipulative processes involved. We may also actually affect the material itself but many through loss by erosion which is irretrievable in most situations over a human timescale.

The discussion thus focuses on a dialogue between desirable states, ease of manipulation (and consequential degree of effort and cost in relation to the returns), intentional and unintentional effects and influences upon the manipu-lations. Thus practices like the addition of lime, organic matter and fertilizer improve soil conditions relevant to crop growth but it must always be remem-bered that these are enacted within an economic context and also that there may be consequences in the wider environment and for the essential non-renewable mineral fraction of the soil. The environmental consequences we consider below and the relationship between management actions and the resource are con-sidered in Chapter 9.

In summary Kirschbaum (1966), Table 7.7 and Bouwman (1990), Table 7.8 have made observations, predictions and recommendations concerning the sensitivity of soil properties and processes to environmental change.

It is also worth considering that human manipulations of soils may have had and have in the future a greater impact on the soil and environment than any

Table 7.7 Soils and climatic change (after Kirschbaum, 1966)

- Focus on soil water carbon, nitrogen and organisms
- The great bulk of soil organic matter decomposes within a few years while some may remain for hundreds of years
- Increasing water status slows down the rates of organic matter decomposition
- Increasing temperature will tend to increase decomposition rates
- Soil carbon storage is liable to decrease – more CO_2
- Methane (CH_4) might also be produced, especially from colder, high latitude organic soils
- Soil warming may lead to increased mineralization of nitrogen (from organic matter to nitrate and nitrogen gases)
- Organisms may also be affected but it is not possible to predict the effect on biodiversity

Table 7.8 Soils and climatic change (after Bouwman, 1990)

- Decomposition is affected by the carbon:nitrogen ratio; increasing carbon will tend to inhibit litter decomposition
- Reduction of permafrost is liable to lead to methane, N_2O and CO_2 production
- Slash and burn activities should be reduced and forestry, especially agroforestry, practices should be increased

changes in climate. Currently there is more of a focus on the ways in which global warming may have an effect on soils with feedbacks to the environment such as the melting of permafrost or increased carbon dioxide leading to enhanced production of methane in wetlands (Hutchin *et al.*, 1996) and the further release of the greenhouse gas methane with a consequent feedback effect reinforcing global warming. There is also the important point that in the geological past, soil processes are thought to have had an important influence on atmospheric composition (Rettalak, 1996) and cannot be viewed as a passive receptor of climatic effects. Additionally, enhanced decomposition of organic matter is envisaged (with temperature rise if moisture is not a limiting factor and greater organic activity) leading to increased carbon dioxide release from soils, again with a reinforcing effect on global warming. Such concerns may detract from concerns about current human manipulations of soils. Should we be more concerned with such projections or should we focus on the human impact on our existing legacy of soil resources? The answer is, of course, that we should be concerned with both, but not to let the former detract from concern about the latter.

Further reading

*	Important reading Recommended reading

Birkeland, P.W. (1984) *Pedology, Weathering and Geomorphological Research.* Oxford University Press.

* Bouwman, A.F. (ed.) (1990) *Soils and the Greenhouse Effect.* Wiley.

Ellis, S. and Mellor, A. (1995) *Soils and Environment.* Routledge

* Kirschbaum, M.U.F. (ed.) (1996) Ecophysiological, ecological and soil processes in terrestrial ecosystems. In Watson, R.T. *et al.*, *Climate Change 1995: II Impacts, adaptations and mitigation of climate change. Scientific and Technical Analyses.* Cambridge University Press/IPCC, 59–74.

Retallack, G.J. (1990) *Soils of the Past: An introduction to paleopedology.* Unwin Hyman.

Trudgill, S.T. (1988) *Soil and Vegetation Systems.* Oxford University Press.

Part III

Productive ecosystems

Chapter 8

Domesticated and managed ecosystems

Summary of key points:

1. There are many prescriptions for land use management in sustainable contexts.
2. Many involve participatory involvement.
3. However, demonstration of benefit is likely to be the most successful way forward.

> ... what is wrong with resource management – it tries to adapt natural systems to the constraints imposed by an artificial economic system. A more sustainable system of management would try to adapt the economic system to the constraints of natural systems.
>
> Jordan, C.F. (1998) *Working with Nature*, Harwood

8.1 Domesticated and managed landscapes

The land provides us with over 90% of all human food, livestock, feed, fibre and fuel (Hurni, 1998). Domesticated landscapes are specifically those lands where the imprint of human cultural activity is dominant. They include agricultural ecosystems (including arable, grassland, horticulture and agroforestry) and plantation woodlands (with no varied structure, tending to be of one age). Domestic gardens are included with urban landscapes (Chapter 5). The main purpose of domesticated landscapes is the growth of food and other ecosystem products of utility. Fabricated, or built, ecosystems are dependent on domesticated ecosystems for food and other products.

Managed ecosystems are also the result of human cultural preferences and activities. They are generally places where the impact is more evident than in wilderness areas (Chapter 5), but where the impact of human cultural energy is less evident than in the domesticated, agricultural systems, where plants and animals are deliberately chosen and cultivated. Intervention might be evident in the managed ecosystem in terms of the production of a commodity such as timber. Here, we include both unplanted forests from which ecosystem products are extracted and also plantation forests which are managed. In the latter case the area is not managed intensively enough to be termed domesticated and although tree species are deliberately chosen and cultivated, there is little intervention between the times of planting and harvesting apart from some trimming and thinning. Here, then, there is a greater element of uncontrolled establishment of other plant and animal species that would be the case in a more intensively managed domesticated forest systems such as agroforestry.

Domesticated ecosystems are concerned with productivity, civilization and culture. In an agricultural context, food security is a clear general issue (Downing, 1996) but predictions of changed growth in a changing climate vary. Singh *et al.* (1998) predict that yields may increase, for example corn and sorghum by 20%, or decrease, for example, wheat and soybean by 20% to 30%. However, they stress the importance of the multiplicity of factors and their interaction, including not only carbon dioxide and temperature but also season length, agricultural zone, crop strain, moisture stress, while Goudriaan and Zadoks (1995) also stress the importance of changed patterns of the effects of pests and diseases as well as weeds. In a complex and uncertain situation the authors wisely make many precautionary remarks about jumping to conclusions. In a discussion of climatic change, carbon dioxide and agriculture, Warrick *et al.* (1986) stressed the significance of different types of photosynthetic pathways possessed by plants, with some C_3 plants being perhaps more sensitive to warming than C_4 (see Inset 8.1 and Figure 8.1).

Inset 8.1   The C_3 and C_4 plant groups (adapted from Houghton, 1996)

During photosynthesis, carbon dioxide from the atmosphere is combined with water to make simple carbohydrates as plant food.

C_3 C_3 species are the majority of plant species globally especially in cooler and wetter habitats. They include all important tree species and most crop species such as wheat, rice, barley, cassava and potatoes. They are called C_3 because the photosynthetic mechanism involves carbon dioxide first becoming involved with a compound which has three carbon atoms. Optimum temperature (°C) ranges are:

Wheat   17–23 (min. 0, max. 30–35)
Rice   25–30 (min. 7–12, max. 35–38)
Potato   15–20 (min. 5–10, max. 32–37)
Soybean   15–20 (min. 0, max. 35)

Rises in temperature could thus affect growth in areas already marginally warm.

C_4 These tend to grow in warmer, more water-limited regions, including many tropical grasses and the crops maize, sugar cane and sorghum. Carbon dioxide is first incorporated into a compound with four carbon atoms. Optimum temperature (°C) ranges:

Maize   25–30 (min. 8–13, max. 32–37)

The optimum range is thus already at a warmer level.

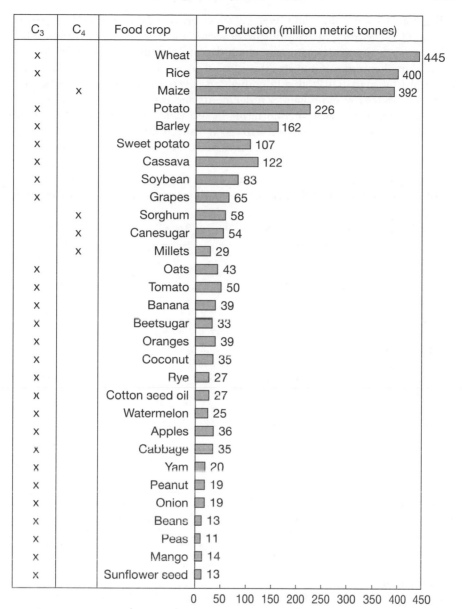

C_3	C_4	Food crop	Production (million metric tonnes)
x		Wheat	445
x		Rice	400
	x	Maize	392
x		Potato	226
x		Barley	162
x		Sweet potato	107
x		Cassava	122
x		Soybean	83
x		Grapes	65
	x	Sorghum	58
	x	Canesugar	54
	x	Millets	29
x		Oats	43
x		Tomato	50
x		Banana	39
x		Beetsugar	33
x		Oranges	39
x		Coconut	35
x		Rye	27
x		Cotton seed oil	27
x		Watermelon	25
x		Apples	36
x		Cabbage	35
x		Yam	20
x		Peanut	19
x		Onion	19
x		Beans	13
x		Peas	11
x		Mango	14
x		Sunflower seed	13

0 50 100 150 200 250 300 350 400 450

Figure 8.1 Agricultural crops: C_3 plants are found in cooler and wetter habitats (see Inset 8.1).

8.2 Managed ecosystems: forests

Of the land surface of the world (variously reported but Hannah *et al.* (1994) give 160 052 691 km²) the World Resources Institute calculated that some 4000 million hectares was covered in forest in the 1980s, with nearly 2900 million as "closed" forest and over 1200 as "open" (Table 8.1). Of the world's forests, 60% are in USSR, South America and North and Central America; only 29% are in Africa and Asia, areas which depend heavily on fuelwood.

Within the topic of forests, the idea of a pristine nature (cf. Chapter 5) disturbed by man is still deeply embedded. Even the putatively dispassionate Table 8.1 above can easily render judgements about loss. "Deforestation" is one the most emotive environmental terms in use today and "Saving the Tropical Rainforest" is seen as a great emblematic venture. Philip Stott (1997), for one, is unforgiving of the folly of these constructs. He writes that notion that the world's forests are natural, and essentially a climatic climax community, is "one of the most persistent, yet pernicious, concepts in world ecology". He continues: "the idea stubbornly permeates nearly everything that is written about

Table 8.1 World forest cover (data from Hannah *et al.*, 1994)

	hectares	%	km^2	*% of world cover
Closed forest				
World	2 859 535 000		28 595 350	17.65
USSR	791 600 000	27.7	7 916 000	4.88
South America	662 505 000	23.2	6 625 050	4.09
North and Central America	528 791 000	18.5	5 287 910	3.26
Asia	431 072 000	15.1	4 310 720	2.66
Africa	221 376 000	7.7	2 213 760	1.37
Europe	137 005 000	4.8	1 370 050	0.85
Oceania	87 186 000	3.0	871 860	0.54
Open forest				
World	1 261 869 000		12 618 690	7.79
Africa	483 943 000	38.4	4 839 430	2.99
North and Central America	277 772 000	22.0	2 777 720	1.71
South America	204 590 000	16.2	2 045 900	1.26
USSR	137 000 000	10.9	1 370 000	0.85
Oceania	71 557 000	5.7	715 570	0.44
Asia	65 120 000	5.2	651 200	0.40
Europe	21 887 000	1.7	218 870	0.14
Closed + open		cum%		
World	4 121 404 000		41 214 040	25.43
USSR	928 600 000	22.5　22.5	9 286 000	5.73
South America	867 095 000	21.0　43.6	8 670 950	5.35
North and Central America	806 563 000	19.6　63.1	8 065 630	4.98
Africa	705 319 000	17.1　80.3	7 053 190	4.35
Asia	496 192 000	12.0　92.3	4 961 920	3.06
Europe	158 892 000	3.9　96.1	1 588 920	0.98
Oceania	158 743 000	3.9　100.0	1 587 430	0.98

World Resources Institute, 1989
* Total world habitat = 162 052 691 km^2

tropical rainforests which, despite so much scientific evidence to the contrary, are still seen by many as the most ancient and undisturbed examples of such climax forest" (a view rehearsed by Park (1992)). Stott writes that there is a powerful ecological "metalanguage" which is in daily use with words and phrases which embody the following: "balance, stable, optimum; age-old, primeval, cathedral; unique, biodiverse, hot-house, jungle, luxuriant" together with "harmony" and the "sin of disturbance" all as descriptors.

By contrast he uses the evidence of palaeobotany to show that tropical rainforests are less than 18 000 years old and that they have changed in composition and extent; many may be younger than some of the world's savannah grasslands; at the end of the last ice age the tropical rainforest occupied an area smaller than they do today. He also cites Whitmore's (1998) *An Introduction to the Tropical Rain Forests*, where far from being in stable equilibrium, the forests are seen as internally dynamic and ever changing with gap phases, building stages and higher stands "ever patterning themselves like a complex amoebic jigsaw" and Stott concludes that "There is simply no climax forest in equilibrium with anything".

It would seem that again the idea of stability seems to reflect human psychological needs rather than any reality. In an interesting juxtaposition to the discussions on wilderness in the last chapter, many landscapes which have replaced forests Stott terms "human landscapes" or "landscapes for humans": "Many of the landscapes which have replaced forests have been wonderful landscapes of hope." He does admit that some replacements have been "landscapes of despair", as also perhaps shown by the study of the environmental transformation of an Appalachian valley by Buckley (1998) but he goes on to say that "this does not mean to say that forests should not be replaced by *positive* human systems" while exhorting tropical forestry to be dynamic, and adaptive, taking change as the norm.

Stott, thus appears to espouse the *ecology for people* notion that we discussed in Chapter 1. I would temper his writings by adding that it seems clear that we *want* to believe that there is something undisturbed and natural and that we should therefore include our spiritual values in our consideration here. Even if we have got it all wrong, this does mean that we should not express a sense of wonder about the tropical rainforest and want to preserve some of it. However, it does seem inescapable that disturbance, far from being a sin, is actually part of the system which gives the forest its characteristics. Studies of gaps (Pickett and White, 1985) show that these are where new growth takes place; fire and windthrow gives new areas for regeneration. There is thus the idea that we might mimic this in management. This does not condone wholesale logging, but we should bear in mind when trying to evolve an "appropriate" attitude, studies such as Johns (1997) who showed that "secondary" or regrowth forest actually supported more species than uncut forests. Indeed, Holmes (1997) published an article in the *New Scientist* supporting the view that logging should actually mimic the patch pattern caused by fires, arguing that forest organisms are already adapted to such a pattern and thus should find it easy to survive in a patch logging approach: "The approach gets you out of playing God. You don't have to make decisions between species, you let the system play out the decisions for you" – and you still have your timber extraction.

"Ecology for people" admits that people have needs of trees for fuelwood and timber in a socio-economic context (Jepma, 1995) and also that people like biodiversity and even the notion of an undisturbed forest. It does not pit people against forests. In these contexts, it is interesting to note that Fairhead and Leach (1996) in *Misreading the African Landscape: Society and ecology in a forest–savannah mosaic*, conclude that far from population rise leading to deforestation, it has lead to the rise of forests round villages and turned fallow areas into woods. Thus, forest increase has already happened, rather than being a novel socio-ecological plan for the future. One might argue that there are "forests" and "forests", new regrowth being seen as less significant than ancient forest. In this context, while Stott argues that the tropical rainforest is "only" 18 000 years old, this is still ancient on a human timescale and might be revered as such. The points here are, I think, that first, people should not be seen as inimicable to forest, second that one might have one's forest and use them (utilitarian argument) and third that old sites have intrinsic value even if our notions of stability and equilibrium are misplaced. Thus, I would agree with Stott that the tropical rainforest is a dynamic, changing, even transitory (on a geological timescale) but these points need not, and perhaps should not, detract from our seeing them as valuable as old. Thus, while I too might have trouble with words such as "stable" and "equilibrium" I do not have the same trouble with "cathedral or unique or, biodiverse, nor perhaps even with "hot-house, jungle and luxuriant". The first set of constructs, from ecosystem science, may mislead us into false assumptions on which to build plans for management and preservation (trying to maintain a stability and equilibrium which is not there, and showing up the sin of disturbance as a rather silly outcome of these constructs); the second set of constructs, however, are cultural ones which embody wonder and reverence. I find these perfectly acceptable as a basis on which to evolve plans for management, conservation and preservation. This perhaps is a general principle – instead of positing pseudo-scientific, "ecological" constructs as reasons for preservation, "why don't we just say we like the idea of having them?" The answer is, of course, immediately that not everyone shares these preferences and that "ecological", scientific or utilitarian arguments (finding medical cures in "untouched" forest) are put forward to counter the activities of those who wish to exploit.

Are we left then with a polarization between the "higher" ground of those who like forests and those who wish to exploit, with only some vague notions of preference and arguments of intrinsic value (which may or may not be contested) to counter them? Arguments about the economic value of preservation are often seen as one way forward (Pearce *et al.*, 1989; 1990) but there is a growing feeling that "*we can have our forest and use it*" (Bruenig, 1996). Exploitation which mimics the inherent disturbances is appealing in this context. Conservation where people reap the benefit is another. The distinction that I would make is that we can meet human needs for fuelwood and timber and that we can do this while keeping our notions of integrity, history and biodiversity, often by the zonation of uses.

Two other threads are evident here. One is from a Sunday newspaper advertisement. "Come and see the Amazon rainforest – one of the world's most ecologically fascinating places" (with comfortable hotels and safety of course). Notwithstanding the arguments about people affecting or even destroying the things they come to see, eco-tourism, if appropriately managed can avoid or minimize such damage, and can boost economies and feed our constructs of wonder about the natural world. Conservation might thus make economic sense.

The second is that wood supplies the main source of domestic energy for more than half the population of the world (Fraser, 1988) (Table 8.2 and Figure 8.2). Fuelwood use is not only influenced by the availability of the wood itself but also is intimately linked with the socio-economic state of a society with, especially, factors like labour availability being important (Brouwer *et al.*, 1997). However, subsistence households are often seen as a leading source of deforestation (Amacher *et al.*, 1996). Surely the way forward here is of local, renewable resources, rather that increasing deforestation? Referring back to Chapter 1, my own experience of Ethiopia was of women facing a daily five-hour trek to gather firewood. This situation was changed by local plantations around the village of Dalocha of eucalyptus which could be harvested for

Table 8.2 Fuelwood as a percentage of total energy consumption (estimated %)

Nepal	98
Chad	94
Tanzania	94
Burkino Faso	94
Ethiopia	93
Somalia	90
Nigeria	82
Madagascar	80
Sierre Leone	76
Ghana	74
Kenya	70
Thailand	63
Sri Lanka	55
Ivory Coast	46
Papua New Guinea	39
Pakistan	37
India	36
Zambia	35
Brazil	33
Zimbabwe	28
Chile	16
Malaysia	8

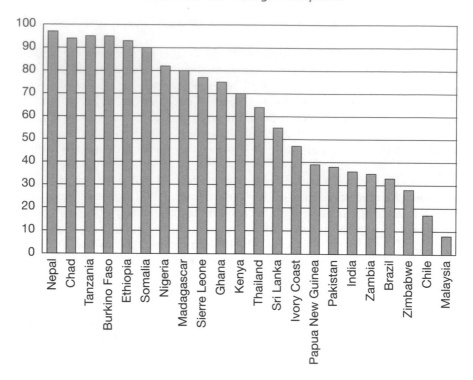

Figure 8.2 Woodfuel use as a percentage of total energy consumption.

firewood on a renewable basis and also released the women for other activities, including healthcare. This also needed some form of social organization in terms of community access to the local resource. This aspect of forest management is stressed by several authors, including, for example, Jewitt (1995) writing about forest protection in India. In Bihar, antagonistic "official" policies have been replaced by joint forest management programmes which involve local people more. Village protection and management committees, together with charismatic village leaders, are important in overcoming tensions over resource use and in promoting successful forms of community action.

Wilson and Bryant (1997) in their study stress the need for acknowledging the relationships between all levels of social organization from the global/international, through the national and to the local. On a broader, national, scale, Didia (1997) in a study of some 55 tropical countries concluded that as democracy increases, the rate of deforestation decreases. The democracy index is a compound of legislative selection, effectiveness, competitiveness of the nomination procedure, party legitimacy and party competitiveness (Table 8.3, Figure 8.3). This does not necessarily mean that democracy is synonymous with local participatory programmes but it may be suggested that the former might facilitate the latter and the latter might improve forest management. In a study using a range of indices, Mather and Needle (1999) concluded that the richest and most democratic countries are characterized by stable or expanding forest while poor and despotic countries tend to experience rapid forest loss. While acknowledging that forest decline may be indicated in the early stages of development, this improves at a later stage of development, their findings are taken to indicate that a high level of development is beneficial rather than detrimental to the sustainability of forests.

Clearly any index used, such as the Democracy Index, and GDP, might be contested and the veracity of the conclusions of these two studies can be judged with reference to Figures 8.3 and for the Democracy Index Study, which are not overwhelmingly convincing, and making comparison with Figure 8.3(b) where it is clear that both negative and positive forest change (loss and gain) are evident at low GDP but that forest loss is clustered around low GDP while forest gain is almost exclusively coincident with high GDP. The existence of forest gain coincident with low GDP again might suggest that the initial stages of development may give rise to deforestation. The authors go on to suggest that forest protection or growth might also be enhanced by a multiparty political system, higher levels of political rights, civil liberties and of human freedom. High correlations could, of course, be coincidental or relate to some other factor, such as richer countries being able to import timber and thus protect their own, but while such additional or underlying factors might be true in some cases, such possibilities are generally minimized in the conclusion that development should be viewed as a positive influence on sustainability.

These findings could be interpreted in a number of ways. First, they refute the myth that development is bad for forests in that there can be a fear that as countries develop, they use more of their resources. This would appear to be true in the initial phases, but why then does forest management improve as development proceeds? Also, does it imply that in order to improve forest cover, development should be encouraged. Mather and Needle, indeed, write that this "does not mean, of course, that everything that happens in the course of development necessarily leads to favourable (forest) trends, or that the sprint for development will ensure forest stability". They find that "the flourishing of civil society under democratic climates may be an important factor as it can bring pressure to bear both on the state and on economic actors through protests, campaigns and boycotts", together with "an expansion of democratic institutions and an opening of bureaucratic decision-making to public participation". Additionally, "even if public opinion is ill-informed, ambivalent or complacent, democratic governments may be accessible to functional elites such as the scientific community or landowners, and hence may be persuaded to take effective action on matters of forest policy". Their fundamental conclusion is that "the achievement of the welfare of nature does not necessarily involve the sacrifice or surrender of human welfare. In essence, ... the conditions that are favourable for humans are also favourable for forest" or, as we said in Chapter 1, ecology for people.

What, then, however, of the countries with a low GDP and high rates of deforestation? Is the conclusion that development would then seem to be

Table 8.3 Deforestation and democracy (data from Mather and Needle, 1999)

Country	Deforestation 1000 ha, ranking	Democracy index	Country	Deforestation 1000 ha	Democracy index, ranking
Brazil	2530.0	13.2	Costa Rica	65.0	20.0
Colombia	890.0	15.0	India	147.0	19.6
Indonesia	620.0	8.0	Venezuela	245.0	19.0
Mexico	615.0	14.6	Jamaica	2.0	19.0
Côte D'Ivoire	510.0	6.0	Trinidad-Tobago	0.8	19.0
Sudan	504.0	5.4	Ecuador	340.0	16.0
Nigeria	400.0	9.4	Mauritius	−0.1	16.0
Thailand	379.0	11.0	Colombia	890.0	15.0
Zaire	370.0	6.0	Peru	270.0	15.0
Ecuador	340.0	16.0	Zimbabwe	80.0	15.0
Peru	270.0	15.0	Sri Lanka	58.2	15.0
Malyasia	255.0	10.8	Botswana	20.0	15.0
Venezuela	245.0	19.0	Dominican		
Paraguay	212.0	8.0	Republic	4.0	15.0
Madagascar	156.0	9.0	Mexico	615.0	14.6
Malawi	150.0	5.0	Bolivia	117.2	13.4
India	147.0	19.6	Brazil	1530.0	13.2
Tanzania	130.0	12.0	Honduras	90.0	12.6
Nicaragua	121.0	4.4	Tanzania	130.0	12.0
Bolivia	117.2	13.4	El Salvador	4.5	11.8
Cameroon	110.0	6.0	Thailand	379.0	11.0
Philippines	92.0	6.4	Malaysia	255.0	10.8
Guatemala	90.0	9.2	Kenya	39.0	9.6
Honduras	90.0	12.6	Nigeria	400.0	9.4
Ethiopia	88.0	3.0	Guatemala	90.0	9.2
Burkina Faso	80.0	3.0	Madagascar	156.0	9.0
Zimbabwe	80.0	15.0	Indonesia	620.0	8.0
Ghana	72.0	3.0	Paraguay	212.0	8.0
Zambia	70.0	6.0	Senegal	50.0	8.0
Benin	67.2	6.0	Mali	36.0	7.2
Niger	67.0	4.2	Philippines	92.0	6.4
Costa Rica	65.0	20.0	Panama	36.0	6.4
Sri Lanka	58.2	15.0	Gabon	15.0	6.4
Central African			Côte D'Ivoire	510.0	6.0
Republic	55.0	4.0	Zaire	370.0	6.0
Senegal	50.0	8.0	Cameroon	110.0	6.0
Liberia	46.0	3.6	Zambia	70.0	6.0
Kenya	39.0	9.6	Benin	67.2	6.0
Mali	36.0	7.2	Togo	12.1	6.0
Panama	36.0	6.4	Sierra Leone	6.0	6.0
Congo	22.0	5.0	Rwanda	5.2	6.0
Botswana	20.0	15.0	Sudan	504.0	5.4
Gabon	15.0	6.4	Burundi	1.1	5.4
Mauritania	13.3	3.0	Malawi	150.0	5.0

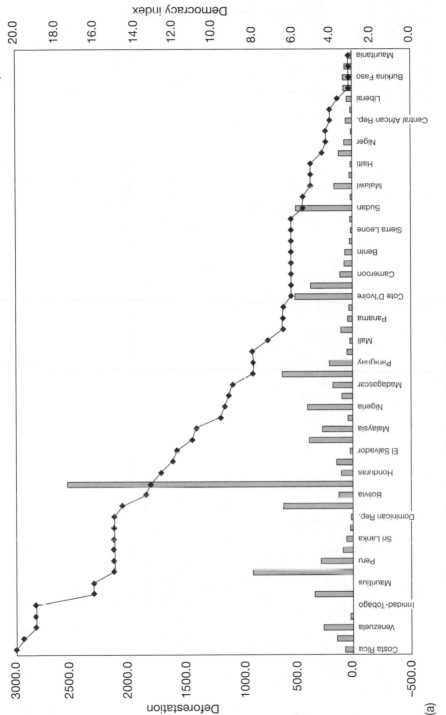

Figure 8.3 Deforestation and democracy (from data of Mather and Needle, 1999).

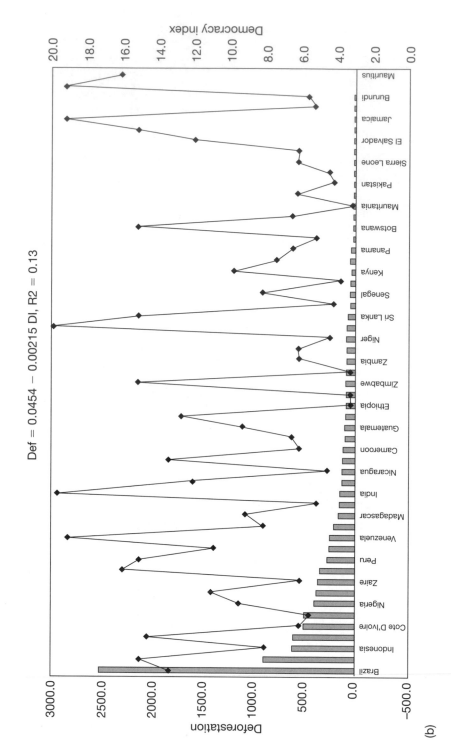

Figure 8.3 *(continued)*

Table 8.3 (continued)

Country	Deforestation 1000 ha, ranking	Democracy index	Country	Deforestation 1000 ha	Democracy index, ranking
Togo	12.1	6.0	Congo	22.0	5.0
Pakistan	9.0	4.0	Haiti	1.8	5.0
Bangladesh	8.0	4.2	Nicaragua	121.0	4.4
Sierra Leone	6.0	6.0	Niger	67.0	4.2
Rwanda	5.2	6.0	Bangladesh	8.0	4.2
El Salvador	4.5	11.8	Central African		
Dominican			Republic	55.0	4.0
Republic	4.0	15.0	Pakistan	9.0	4.0
Jamaica	2.0	19.0	Liberia	46.0	3.6
Haiti	1.8	5.0	Ethiopia	88.0	3.0
Burundi	1.1	5.4	Burkina Faso	80.0	3.0
Trinidad-Tobago	0.8	19.0	Ghana	72.0	3.0
Mauritius	−0.1	16.0	Mauritania	13.3	3.0

needed in order to protect forests a naïve one? Many texts actually favour a participatory approach (which may or may not be the preserve of democratic countries), and in *Who will Save the Forests?* Banuri and Marglin (1993) stress the importance of the practices of "age-old rural communities ... which have ... managed the environment in [a] sustainable fashion down the centuries" (Plate 8.1) and that the development strategy should be "one that restores the autonomy of local arrangements". Similarly, Ghai and Vivian (1912) stress the importance of "*Grassroots Environmental Action: People's participation in sustainable development*". Certainly, the history of forest management in India has shown that social forestry has great potential, but also can have its own pitfalls. Writing in *The Greening of India*, Vohra (1985) slams the centralized organization of forest use with the "barbed-wire and forest guard approach" with forest protection and the exclusion of people saying that it is "a ridiculous proposition that a quarter of the country's total land resources ... should be allowed to be managed by a handful of State forest departments ... How utterly ridiculous this proposition is will be apparent if we imagine for a moment what would have happened to our agriculture if our agricultural lands had been under the management of a few agriculture departments rather than under the millions of hard working farmers with a personal stake in the productivity of the soil". Cernea (1988) charts how the history of social forestry which was intended to redress such complaints has not been an unequivocal success and even uses a sub-title "The anatomy of failure", focusing on difficulties of land tenure and the appropriate nature and size of the social organization involved. The lessons learnt in the early days of simply planting for the community were that there are three important social variables to be considered:

Plate 8.1 Ladakh. Banuri and Marglin (1993) stress the importance of the practices of "age-old rural communities ... which have ... managed the environment in (a) sustainable fashion down the centuries".

1. complex land tenure systems and processes affecting it (social forestry) at deep levels;
2. the community unit with its internal interactions, non-homogeneous groups and inability to act consensually; and
3. the behavioural patterns of individual farmers.

The more recent studies in the literature stress the importance of multiple stakeholders, multiple products, species diversity, multiple time horizons (different seasons), natural regeneration, site specific planting, local knowledge and participatory planning (e.g., Hobley, 1996). Krott (1996) stresses the importance of self-regulation forest policy, with "all those with a stake in forestry expect(ing) forest policy makers to defend their (the stakeholders) position against exaggerated demands and attempts at control from outside" while Beese (1996) stresses multifunctional use with habitat, production and cultural/social functions all developed in an integrated assessment. Beckley and

Korber (1995) stressed sociological analyses of stakeholder values, social movements, conflict and its resolution, poverty in forest-dependent regions, and complex organizations in the practical applications of sociology in a forest context. In terms of communal forest management, Dejong *et al.* (1995) and Sharma (1993) conclude that such schemes must be sociologically viable for each region while Dove (1995) shows that a lack of sociological perspective leads to a tendency to add resources perceived to be in short supply instead of removing institutional obstacles. The preferences of villagers for traditional timber species and the importance of providing communities with a share of the income from the sale of forest products is emphasized by Chatterjee (1995) and the linking of land use practices to people's social values and needs is stressed by Ffolliott *et al.* (1995), as is the maximization of multiple benefits including wood for fuel, poles and posts and building materials, fruit and nut production, fodder, browse and forage; the improvement of local microclimate for improved agricultural crop production and protection of soil from erosion while Pelos (1993) recognizes the importance of negotiation and re-negotiation in terms of traditions of forest control in Java. Finally, Wiersum (1995) reflects on "*200 years of sustainability in forestry – lessons from history*" and draws three lessons:

1. the need to recognize the different nature of ecological limits and social dynamics;
2. the role of dynamic social values with respect to forest resources; and
3. the significance of operational experiences in trying to attain sustainability within a concrete context.

While Tinker (1994) was able to write of "exaggerated expectations of the potential of community forests" it is, however, clear that "simple solutions to single problems may actually create problems" (Rocheleau, 1995) and that the sensible way forward is some kind of participatory forestry which maximizes the social benefit through multi-use, providing timber, fodder and fuelwood (Timberlake *et al.* (1993) Chapter 9, and Vira *et al.* (1998)). In attempting to implement such schemes, there has clearly been a learning curve which as Honadle (1993) writes in *Defining Sustainable Forestry*, "attitude and education, while important, are not enough". Ahlback (1995) stresses the importance of technology, resources, motivation and institutions, the latter showing that there needs to be institutional changes, as O'Riordan and Jordan (1999) stress and which important theme is discussed in the concluding chapter of this book. The underlying "win-win" is that diversity of species could coincide with diversity of useful products and also increase the capacity to cope with change.

8.3 Agricultural ecosystems and the environment

Agricultural activity depends upon and affects the soil and also it affects the wider ecosystem. The primary core concern of agriculture is productivity and the relationship between land quality, inputs (seeds, fertilizers, pesticides) and output (Figure 8.4). There may also be a secondary consideration in terms of the effects of agricultural production on the soil (Figure 8.5). There are

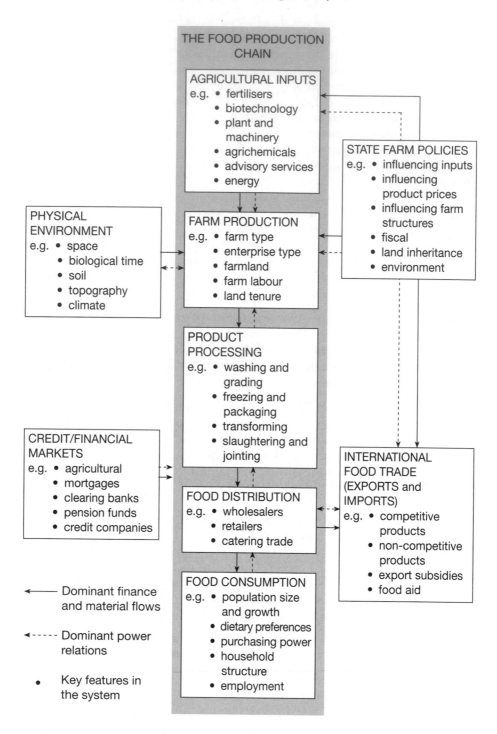

Figure 8.4 The agricultural system (adapted from Bowler, 1996).

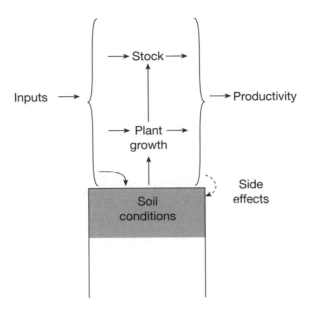

Figure 8.5 Productivity in agriculture.

also wider considerations (Figure 8.6) which admit that agriculture shapes the landscape and affects the wider environment (see Tinker, 1988; Briggs and Courtney, 1985). A fundamental aspect is the change of habitats caused by the existence of agriculture which, apart from the maintenance of rough grazing, inevitably means the clearance or at least the substantial alteration of existing vegetation. A further major way in which these wider effects are felt is through water which runs off agricultural land: this may transport sediment, agricultural chemicals and organic matter from fields (Winteringham, 1985). Emissions to the atmosphere may also be involved. We can thus think of on-site and off-site effects. There is no doubt, however, that an agricultural area can be viewed as an ecosystem, with flows of energy, water and nutrients through the system and that, although it is managed on a per field basis, these flows proceed irrespective of field boundaries (Figure 8.7).

Issues which have involved agriculture and the environment include:

- the use of fertilizers applied to the soil but which have also then entered watercourses, leading to eutrophication;
- the use of pesticides and herbicides which have found their way into water, air and non-target organisms;
- cultivation and the degradation of soil structure, especially in relation to the use of heavy machinery on the land, with consequent effects on plant growth and water runoff;
- soil loss through erosion, including both on-site and off-site impacts such as downstream silting;
- general considerations of conservation of wildlife on farmland.

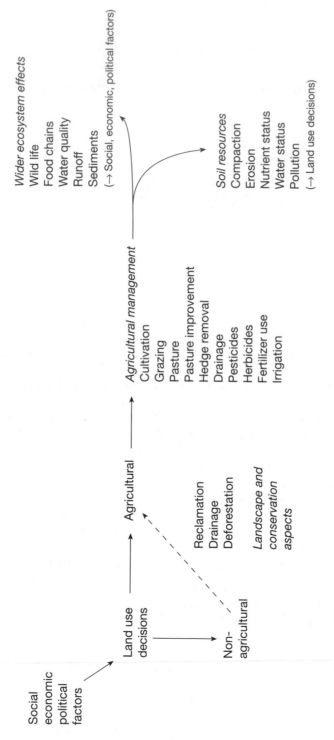

Figure 8.6 Wider consideration in agriculture.

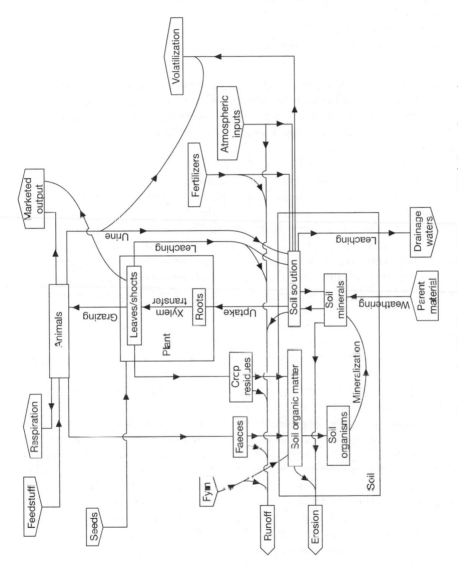

Figure 8.7 The agricultural ecosystem (modified from Briggs and Courtney, 1985). Fym = farmyard manure.

The responses to these issues have included:

- technological – improved techniques, machinery;
- timing – ensuring that the timing is appropriate;
- placement – ensuring that the spatial distribution of material is appropriate;
- the judicious use of fertilizers, pesticides and herbicides;
- curbs and bans;
- a call for organic farming;
- codes of good agricultural practice;
- integrated crop management.

It is also true that many policy changes have been made, often in relation to food supply or other political motivation which have not necessarily had an explicit environmental motivation but which nevertheless have had an environmental impact. There are also specific environmental policies designed to improve the environment.

In many ways, agricultural activity has come to be seen as hostile to wildlife and the environment, as encapsulated in the book by Coleman-Cooke (1965) with the graphic title: *The Harvest That Kills*. High productivity can be equated with intensive farming practices such as the high usage of pesticides and fertilizers and the use of large fields without hedges or woodland. However, since people now have the awareness that we both need food and value a pleasant landscape, recent endeavours have attempted to optimize productivity while promoting wildlife and enhancing the environment (e.g., Soper and Carter (1985) in *Modern Farming and the Countryside*). This may be achieved by altering land use type but more realistically by modifying land use practice, often, but not necessarily, in cognizance of indigenous practices (Reij *et al.*, 1996) and with a dialogue between indigenous practices and "modern methods", as discussed further, below. Here, the crop is still grown – as is viable in an economic context – but with attention to timing or spatial arrangement of the management practice any detrimental effect on the soil or wider environment can be minimized.

The production ethic has dominated agriculture (see, for example, Cooke (1983) *Fertilising for Maximum Yields*). This continues to be true in many parts of the world, though in Europe, North America and some other developed regions, surpluses and the need to conserve the soil and wildlife have played an increasingly important role in the last fifty years. In the UK the Second World War played an especial part in British attitudes to agriculture. With blockades at sea, imports became scarce and the island was thrown back on its own agricultural resources to feed the nation. Reclamation of "waste" land became the order of the day through a "ploughing up" campaign – agricultural productivity became synonymous with patriotism.

Even though the war did not end till 1945, E. John Russell was writing from Rothamsted Experimental Station "Agriculture in Europe after the war: What can we do?" in the *Journal of the Royal Agricultural Society of England* for 1943. A technical Advisory Committee was set up in 1942 with representatives from the US as well as the UK. Necessities were identified such as seed, implements, man-power, livestock and fertilizer. Before the war, Europe was self-sufficient in nitrogen fertilizers (by synthesis in factories) but phosphate was imported; during the war the supply of the latter was cut off and also a reduction in livestock numbers led to a corresponding drop in farmyard manure available. Russell wrote: "After the War it will be essential to increase the productiveness of the soil and the most rapid treatment will be to use nitrogenous fertilisers. Phosphatic fertilisers must be supplied to make up for the deficiency accumulating over the War years. Fodder crops and grass and clover will be vitally important in restoring soil fertility and must be increased as speedily as possible."

The need for self-sufficiency and productivity were very clear, together with a decreased reliance on imports. It is therefore small wonder that agriculture saw a rapid upturn in agro-chemical use and associated rises in productivity. This ethic still lingered into the 1990s when set-aside policy, introduced to reduce European surpluses, took land out of production but did not turn it over to wildlife conservation (a missed opportunity, according to green pressure groups) because of the idea that it still had to be ready to be returned to agriculture if the need arose.

Farming in the UK seems to have changed more in the last thirty or so years than at any time before: other changes, even agricultural revolutions such as mechanization, all were underwritten by the production ethic. Now the influences are concerned as much with environment, resource sustainability, food quality as much as quantity and interest in wildlife conservation is such that it cannot be ignored as a substantial part of the agenda. In addition subsidies for production have at least been questioned if not diminished or withdrawn and as a result whole ways of lives are altering and agriculture is becoming as much about sustainability, "the countryside", regional identity, tourism and leisure as about production.

8.4 Sustainable agriculture

A basic view of sustainable agriculture is one which does not impair the capacity of the soil (and in the case of grassland, the vegetation) to produce the desired product in the long term; a broader ecological view takes into account the whole process, including the manufacture of substances applied to the land and any off-site effects, other definitions also include economic and social sustainability (Edwards et al., 1990; Ilbery et al., 1997; UK Round Table on Sustainable Development, 1998). In introducing Edwards et al.'s book on sustainable agriculture, Harwood (1990) defined it as "an agriculture that can evolve indefinitely toward greater human utility, greater efficiency of resource use, and a balance with the environment that is favourable both to humans and to most other species". Sustaining the soil resource is now also seen as central (Reij et al., 1996). One of the biggest impediments to sustainable use is land tenure (Bandyopadhyay 1996) and indeed Eckholm, (1979) made a direct link between the monopolization of "the best land by a few" while the dispossessed "try to eke out a living on lands, that for ecological reasons, ought not to be farmed" – and he cites steep hillsides, desert fringes and slash and burn in tropical forests as cases in point, in other words where agriculture is not only economically marginal but also ecologically unsustainable.

The OECD (Organization for Economic Co-operation and Development) (1995) realized that there were multiple definitions of sustainable development. They thought that central to the ideal of sustainability was the definition of Repetto (1986): "a development strategy that manages all assets, natural resources, and human resources, as well as financial assets, for increasing long-term wealth and well-being". They stress that the asset value of a natural resource, such as cropland, is not just concerned with the economic value of the crop. Their criteria are:

1. marketed goods – the value of the crop;
2. use values – which derives from the contribution the resource makes to important natural processes (e.g. storing rainwater);
3. direct enjoyment (e.g. walking in the countryside); and
4. existence value – reflecting the intrinsic worth of the resource independent of its value in use.

Most definitions are formulated in rather broad and abstract terms and this often makes for difficulties when trying to put them into practice. However, the OECD actually attempt this and list examples of agricultural practices that have a high potential for sustainability:

1. Conservation tillage – systems of no or minimum tillage which reduces the amount of soil disturbance and so lessen runoff and loss of sediments and nutrients.
2. Rotations – growth of different crops in succession. These may either exploit different aspects of the soil resource or indeed improve the soil especially if green manuring is involved where a crop, especially a nitrogen fixing one, is ploughed in.
3. Intecropping – again different crops are used in a spatial rather than a temporal sequence; this may include agroforestry where the plants not only exploit different aspects of the soil resource but also provide different benefits.
4. Silvipasture – trees, grassland supporting mixed livestock.
5. Precision management – detailed measurement of soil conditions, involving soil testing and only applying what is deficient; similarly the sparing or judicial use of pesticides only when required in an integrated pest management which may also involve biological control, especially measures which increase the numbers of pest predators.

It is interesting to compare these definitions of sustainable land use with the claims for organic farming below. Agroforestry is a general term applied to "land-use systems in which trees or shrubs are grown in association with agricultural crops, pastures or livestock – and in which there are both ecological and economic interactions between the trees and other components" (Young, 1989, p. 11). The benefits include the build-up or organic matter and nutrients, shelter and moisture conservation beneficial to the crop and the parallel production of fodder, browse and timber from the tree. Species selection and spatial arrangement is important in order to maximize the benefit and the selection of species, such as *Leucaena* which fix nitrogen can be particularly useful. Many claims have been made about the potential and the success of agroforestry and forest farming (Douglas and Hart, 1984) but studies show that a simple implementation is not necessarily enough. This is because other factors like topography and soil type have to be allowed for as well as farm size, farmer's experience and education plus the economic context are all important when trying to assess measures of success (Pattanayak and Mercer, 1998) and that it is not automatically profitable and that labour availability has to be carefully considered (Jepma, 1995, pp. 131–136).

Thus the many definitions and general desirability of sustainability are evident but the methods and successes of implementation are less clear. This endeavour is very much one which highlights the dialogues between the practitioners on the one hand and the forces acting upon them on the other. Agriculture is a blend of demands for products, economic contexts, market forces and viability, agricultural policies, institutional influences and the availability of crop types, animal breeds and inputs such as fertilizers and other additives. Also involved are cultural heritages, motivations and expectations both about agricultural methods and the type of produce as well as the amount and type of labour involved. This is together with environmental and ecological factors concerned with the weather and climate, topography, soil type and condition and the practitioner's knowledge and skills concerning the plants, animals and soil as well as technical expertise. Any day-to-day decision about what, when and where to grow or raise is going to be a synthesis by the practitioner of many if not all of these factors.

Many see the problem facing sustainable agriculture lies in terms of a polarization between the practitioner – the "local people" – and the institutional forces (Stocking, 1998) with the answer lying in greater interactive participation and self-mobilization of the local people in decision making (Pretty, 1998). Clearly this can provide a central theme but the list of factors above means that this will be a complex matter in practice.

In terms of polarization between local people and institutions, Stocking (1998) writes of how professionals tend to perceive local society as one homogenous group and can be also uncomfortable about their precarious base in their own institutions. Both of these have to be addressed in terms of increased perceptions of diversity in a client group and greater security. The way ahead surely involves a "third way" between the polarized views outlined in Table 8.4 or at least a dialogue between them.

All of these polarizations can be dealt with more intelligently. Essentially, in terms of local people, this involves respecting local attitudes in terms of the legitimacy of social and cultural norms (see Berkes (1999)), the need for land tenure, the importance of local empowerment, respecting traditional knowledge and getting participation right on the one hand and education about degradation and soil protection on the other. For institutions, this involves recognition

Table 8.4 Polarized stereotypical views of local people and institutions (adapted from Stocking, 1998)

Local people	Institutional view (from soil science agronomy/professionals)
Soil – supports livelihoods	Needs protecting
Soil degradation – not perceived	Urgent attention needed
Social and cultural norms an essential consideration	Can be seen as a nuisance
Land tenure essential for investment	A distant political issue
Stability and political commitment – nice to have the support	Concerned with stability and salary within own institution
Family income for labour essential	Get them working, not lounging
New technologies – risky for me	Of course it is good
Local empowerment – feel not trusted	Not to be trusted
Minimize subsidies – no we need paying for conservation	Minimize subsidies
Participation – asked then ignored	Makes job more difficult
Different institutions contradict	Different institutions compete
Traditional knowledge respected	Traditional knowledge second rate
Long-term view	Three-year funded project

of the pressures which competition and short-term contracts bring and a more enlightened attitude towards local cultures, land tenure, labour, empowerment and tradition. Empowerment particularly has to be negotiated as do subsidies and the probable outcomes of new technologies. These aspects, and particularly participation, should be interactive (see Röling and Wagemakers (1998) *Facilitating Sustainable Agriculture: Participatory learning and adaptive management in times of environmental uncertainty*).

In terms of participation, Pretty (1998) provides a spectrum of modes of participation as shown in Table 8.5, of which *Interactive* involves the greatest dialogue and *Self-mobilizing* the greatest empowerment.

Pretty concludes from the analysis of projects in many countries that participation is the most significant factor in contributing to project success, with interactive participation being the key and that it is when people are involved in decision making at all stages that the greatest success is achieved. This means that the label "participation" *per se* is not enough, the type and effectiveness of the participation is critical, interactive and self-mobilized being the most effective.

As a case study, Defoer *et al.* (1998) provide an example of participation in soil fertility management in Mali. Here, planning and evaluation of activities is done on a yearly basis. Farmers and researchers jointly evaluate the merits of different procedures. This resulted in greater recycling of crop residues as litter and fodder, increased use of composting and improved soil fertility by visible results. This was from farmers' own trials involving examples of farms which were representative of a range of different levels of farm management – the

Table 8.5 A typology for participation (adapted from Pretty, 1998)

Type of participation	Description
Manipulative	A pretence
Passive	Told what has been decided
Consultation	Consulted without any necessary involvement in decisions
Bought	Materials provided with no stake in practices
Functional	Forming groups to meet predetermined objectives
Interactive	Joint analysis, development of action plans and groups. Learning methodologies to seek multiple perspectives, shared determination of how resources are used
Self-mobilization	Initiatives independent of external institutions. Contact with institutions for resources and advice but retaining control over resource use

villagers themselves having identified this range and selected examples from it. Practices which were sustainable and enhanced the farming operation were identified and jointly recommended.

It is interesting to compare these more sociologically sensitive analyses with earlier prognoses by soil scientists. Syers and Rimmer (1994) list several pre-requisites for the sustainable use of soil (Table 8.6). All of these are true but I think that the order in which they are presented is interesting with italics showing that the societal context is almost an afterthought when, in fact, the analyses above show that the reverse may be true. This is not to say that the technical points about soil pH, organic matter and so on are not important or can be ignored. Such knowledge is fundamental and thus social scientists cannot claim to have made all the important points – but then neither can soil scientists either. A hybrid physical science–social science approach is what is actually fundamentally important. Technical knowledge has to be demonstrably useful in societal contexts, which is why the paper by Defoer above is important.

If we may now return to the list of factors which may be involved in decision making, we can classify the degree to which they are under the practitioners'

Table 8.6 Aims of sustainable land use systems (Bennett (1994) in Syers and Rimmer (1994))

- Replenish the nutrients removed
- Maintain the soil's physical structure
- Ensure constant or increasing levels of organic matter
- Maintain soil pH
- Avoid the build-up of substances to toxic concentrations, e.g. salts or other contaminants
- Avoid outflows of polluting substances
- Avoid the physical loss of soil particles
- Maintain a desirable balance in the flora and fauna
- *And then* continue to meet the needs of the communities who depend on it for their livelihoods (my emphasis in that I would put this first)

control. It is readily appreciated that economic and environmental factors are outside the control of individual farmers, policies and societies can be subject to influence and technical resources can be mobilized but the domain of individual control lies only really in terms of those practices which may be personally adopted.

Economic
- market forces
- economic contexts
- demands for products

Environmental
- environmental and ecological factors concerned with the weather and climate
- topography, soil type and condition

Policies
- agricultural policies
- institutional influences

Societal
- cultural heritages
- motivations and expectations
- amount and type of labour involved

Technical
- availability of crop types, animal breeds and inputs such as fertilizers and other additives

Personal
- practitioner's knowledge and skills concerning the plants, animals and soil
- technical expertise

Land use practice (see Section 8.6) is thus a critical factor in sustainable farming rather than land use or crop type. Individuals will crop what is needed, what will sell, or what is otherwise indicated through policies with the minimum of environmental modification (unless there are subsidies to do more, e.g. for land draining) and will relate to cultural and societal expectations, labour availability and the availability, risk-benefit perception of technologies and capability for investment in technology, all according to the knowledge and skill which the farmer possesses. Thus in terms of promoting sustainability, what you do (e.g. which crop to grow) can be important but more important is the way in which you do it (e.g. type of tillage, amount of recycling, use of mulches and so on). Sustainability thus means optimizing and sustaining economic viability and the conservation of the soil resource. Increasing soil fertility through recycling crop residues which leads to increased productivity (Defoer *et al.*, 1998) seems to be a prime example. Thus, the win-win situation is when both productivity and conservation are mutually and interactively maximized – soil is conserved for the future and the farmer makes a viable living. Here the incentive is clear – and all the better if the beneficial situation is realized through genuine participatory investigation and demonstration.

However, if sustaining the soil resource seems to lead to a drop in productivity, what then are the incentives? This critical situation is usefully discussed by Giger (1998) in a comparative evaluation of incentives and subsidies. An *incentive* is seen as a broad term which encompasses all motivations which stimulate people to act. A *subsidy* is a specific incentive "used by the state or private actors to reduce the cost of a product or increase the returns from a particular activity". Compensatory payments may also be involved. Subsidies are often politically motivated – for example the post-Second World War drive for productivity in the UK discussed above, and practices have tended to make them anti-ecological, the prime example of which is the drainage subsidy which made agriculture economically possible in what might have otherwise been ecologically valuable areas.

There are also other arguments against subsidies (Giger, 1998). Essentially a permanent subsidy may be very costly, distort management decisions and the market forces as well as making farmers and communities dependent on outsiders. They may also be explicitly grounded in the assumption that people cannot help themselves. However, *temporary conversion subsidies* may well be very worthwhile in that they can cover the cost of short-term losses before long-term benefits accrue. In 1999 a UK scheme for conversion to organic farming is meeting with some considerable interest in this context (see Section 8.5 below). There is a perceptible marked reluctance to acknowledge and adopt soil conservation practices (Unwin, 1998; Currie, 1998; Sanders *et al.*, 1998) and it is clear that some triggering mechanism is needed (Reij, 1998). Also, as well as participatory schemes and personal contacts, such conversion subsidies may well be an important factor.

In summary, there seems to be recognition of many elements of sustainable agriculture and of the need for a dual approach where the relevant knowledge and outcomes are demonstrated in a participatory atmosphere in order to achieve the necessary motivation. There has to be some incentive before change will occur. Land use practice is a critical factor (see Section 8.6).

8.5 Organic farming

Organic farming seeks to avoid the use of manufactured chemical fertilizers and pesticides and to mimic natural ecosystems by returning organic matter to the soil, thus closing the cycle of nutrient fluxes. It tends to yield a lower productivity than intensive farming and a product of more variable quality. It is widely seen as not so environmentally damaging as intensive farming and, indeed, presenting opportunities for wildlife (Chamberlain *et al.*, 1999; Reganold, 1988, 1989; Reganold *et al.*, 1987), though it is true that contamination of water courses can occur from organic manures if applied without due regard for this possibility.

The United States Department of Agriculture defines organic farming as:

> Organic farming is a production system which avoids or largely excludes the use of synthetically compounded fertilisers, pesticides, growth regulators and livestock feed additives. To the maximum extent feasible, organic farming systems rely on crop rotations, crop residues, animal manures, legumes, green manures, off-farm organic

wastes, and aspects of biological pest control to maintain soil productivity and tilth, to supply plant nutrients and to control insects, weeds and other pests.

McCann *et al.* (1997) report on a survey of attitudes of organic and non-organic farmers, finding that both groups shared a concern about the risks of farming, but that the former had a more long-term view, with a greater willingness to incur present risk for long-term conservation benefits. Sullivan *et al.* (1996) also reported not only support for the notion of living in harmony with nature and the land but also more satisfaction with their lives, a greater concern for living ethically and a stronger perception of community.

The principles and practices are expressed in the standards document drawn up by the International Federation of Organic Agricultural Movements:

1. To produce food of high nutritional quality in sufficient quantity.
2. To work with natural systems rather than seeking to dominate them.
3. To encourage and enhance biological cycles within the farming system, involving micro-organisms, soil flora and fauna, plants and animals.
4. To maintain and increase the long-term fertility of soils.
5. To use as far as possible renewable resources in locally organized agricultural systems.
6. To work as much as possible within a closed system with regard to organic matter and nutrient elements.
7. To give all livestock conditions of life that allow them to perform all aspects of their innate behaviour.
8. To avoid all forms of pollution that may result from agricultural techniques.
9. To maintain the genetic diversity of the agricultural system and its surroundings, including the protection of plant and wildlife habitats.
10. To allow agricultural producers an adequate return and satisfaction from their work including a safe working environment.
11. To consider the wider social and ecological impact of the farming system.

Lampkin (1990) maintains that the criticisms of current (non-organic) farming practice are that:

1. It damages soil structure.
2. It damages the environment.
3. It creates potential health hazards in food.
4. It has brought about a reduction in food quality.
5. It is an energy-intensive system.
6. It involves intensive animal production systems which are ethically unacceptable.
7. It is economically costly to society, and increasingly so to the farmer.

All of these points can, of course, be contested. The claims for organic farming might seem to some to be unduly idealistic and the system not necessarily economic – organic farming is often seen as only financially viable because of the organic premium which some people are willing to pay. The criticisms of non-organic farming are not necessarily so. What is interesting is how the two are seen as opposable, even mutually exclusive, when there are elements of both which are worthwhile especially if one asks the question, how can you have an economically viable and productive agriculture with high quality products without damaging the soil or the wider environment?

The answer may lie in a hybrid approach which agrees with all the organic principles above but still maintain the judicial use some agrochemicals. A system of integrated crop management is important here, defined as "establishing resource-conserving practices and the optimization of agricultural input and management with the aim of decreasing costs while maintaining productivity and benefiting the environment (Royal Commission on Environmental Pollution, 1996). This approach coincides with some of the definitions and possible practices which promote sustainable land use above. Pesticides could then be used on occasion, rather than routinely, using target specific and readily decomposable substances. Organic manures could be used routinely but soil nutrient levels would be monitored and topped up with fertilizers where necessary (and even under strict organic farming naturally occurring mineral fertilizers may be used). Cultivation would be timely so as to avoid soil structural damage (Strutt, 1970) and buffer strips next to water courses would be uncultivated and untreated which can help to minimize water pollution from agricultural dressings (Haycock *et al.*, 1997). It is interesting that reports exist of the adoption of such an approach. The findings would appear to be that productivity might be somewhat reduced but far from leading to reduced profitability, the approach is more economically viable because of the savings in expenditure on agrochemicals.

As I write, the new millennium has seen a great upsurge in the interest in organic farming. This is widely seen as nested in food scares connected with BSE and GM products. A few years ago the UK Society for Chemical Industry (SCI) journal *Chemistry and Industry* ran a comment page on "*The folly of organic farming*" (Avery, 1997). Here one of the objections was that lower productivity would lead to more extensive land use and *less* conservation in terms of less available land. Overtaken by events, the SCI called a conference in London on 23 March 1999 where it was reported that as conventional farmers were going bankrupt, the Ministry of Agriculture funded Organic Conversion Information Service was swamped with enquiries about a scheme to subsidize the conversion to organic farming. While much of the interest came from grassland farmers, the interest was clear. Nicholas Lampkin cited the *Farmers Weekly* as moving from the position of "Unless you are keen to live at little more than subsistence level, organic farming is not for you" in 1992 to "Organic farming with realism could become a godsend for the industry" in 1999. Economic opportunities presented in relation to public opinion are clearly very powerful.

8.6 Decisions on land use type and land use practice

We have said that land use practice is probably more important than land use type. A steep slope may be cultivated but the presence or absence of terraces or other conservation measures is critical in determining the amount of soil

---

Inset 8.2 The necessary conditions for effective soil conservation programmes (Napier, 1990)

1. Political development to support action to reduce the economic, social and economic costs of soil erosion.
2. Long-term allocation of human and economic resources by governments.
3. Government agencies with sufficient autonomy to address long-term issues with immunity from short-term political influences.
4. Development of well-trained professional staff in soil conservation agencies.
5. A farm population which is informed on the causes and remedies of soil erosion.
6. Stewardship orientation and attitudes among land operators.
7. National policies which place a high priority on soil conservation.
8. Developing policies for agricultural development and soil conservation which are consistent and complementary.
9. The creation of complementary national environmental programmes.
10. The involvement of both physical and social scientists who will contribute to the creation and implementation of soil conservation policies and programmes.
11. An interdisciplinary professional society committed to the integrity of soil and water resources.
12. Political leadership which will be willing to implement soil conservation policies which some segments of agriculture will find oppressive.

---

erosion. Similarly, the timing of ploughing and the time and season over which the soil is left bare without cover of growing vegetation or a mulch can be critical. In fact these practices are easier to announce than to put into practice in relation to specific socio-cultural contexts. There are many texts which give prescriptions as to how to improve and maintain soil fertility and minimize soil erosion through the use of appropriate land use management techniques, including Syers and Rimmer (1994) mentioned above, Weischet and Caviedes (1993), Morgan (1980; 1986), Greenland and Lal (1979), Greenland (1981) and Young (1989). What is interesting is the dual trend of working with nature (see Jordan (1998) *Working with Nature: Resource management and sustainability*) and the growing awareness of the importance of social factors discussed above (see also Inset 8.2; Figures 8.8, 8.9 and 8.10).

The need for sociological awareness is not new. In the 1950s, Blaut *et al.* (1959) made "A study of the cultural determinants of soil erosion and conservation in the Blue Mountains of Jamaica" which was followed up by Floyd (1970). Extensive soil erosion was seen to stem from a history of continuous cultivation (Plate 8.2) for coffee over long periods of time followed by:

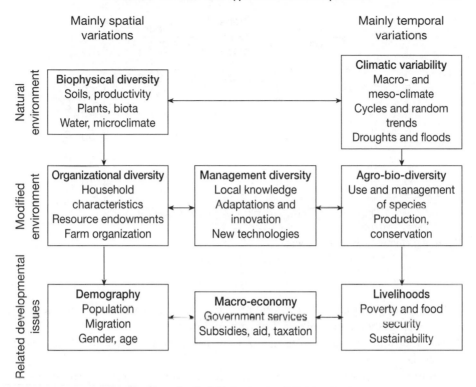

Figure 8.8 Agrodiversity (after Brookfield and Stocking, 1999).

1. clean weeding (especially between bananas);
2. exposure of soil for yam planting at the wettest time of the year;
3. hoeing giving rise to downslope movement;
4. lack of tree crops; and
5. absence of soil conservation measures.

Government recommendations were as follows:

1. grass lines parallel to the contour;
2. ditching;
3. tree crops;
4. use of animals (grass cover and grazing);
5. integrated farm plans – suit crops to the slope, e.g. steep lands with trees.

The latter list seems eminently sensible, easy to announce and concerned to protect the soil resource and also in terms of capacity building. However, it was found that the adoption of appropriate practices was, in fact, very limited and that the reasons were largely socio-cultural:

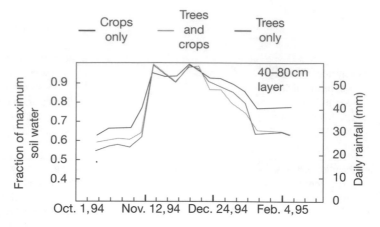

Figure 8.9 Agroforestry and soil water.

1. Community leaders were catholic preachers and lay teachers, not farmers, so any changes were not followed or imitated.
2. Local agricultural extension agents were not around long enough to attain prestige.
3. Community structure – women exerted pressure to grow (clean weeded) cash crops for market (tree crops and bananas were marketed by men). The women derived pleasure from meeting friends at the market and the satisfaction of contributing to the family income. Also there was instability of common law family unions – children were attached to the mother and with the mother dependent on cash crops this represented a safe form of family income. This factor was more important in the low income groups; in higher income groups more people got formally married.
4. Land tenure – the feeling that they will not plant trees on another man's land (even though it might be rented for 30 years) since they would not be sure of reaping the benefit.
5. Capital scarcity: (i) could not afford labour/money for ditching and other soil conservation work. Even with a Farm Development Scheme (FDS) where the government pays a part of the cost, the farmer still cannot afford the labour or the rest of the money (alternatively, this may be seen as an excuse to bolster up prejudice against the proposed measures); (ii) temporary capital could be lost during tree planting and initial growth.
6. Land anxiety – land scarcity meant people were very conservative about anything that may remotely affect their land security and title to the land.
7. Universal latent distrust of the peasant farmers for a government with which they have had little contact.
8. Lack of perception of soil erosion, its causes and consequences or control, or, if perceived, that anything can be done about it.

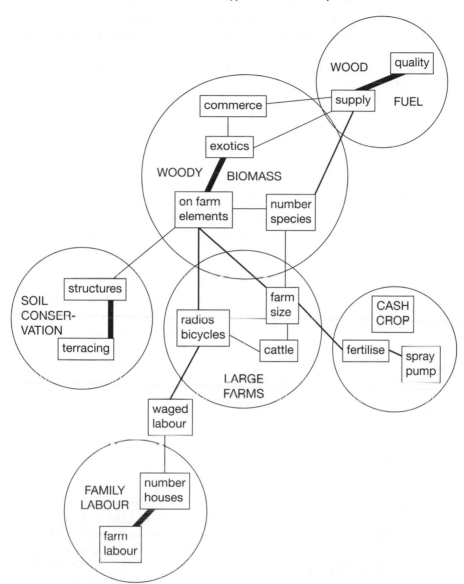

Figure 8.10 Interconnections.

In a survey of practices, there were found to be some unconscious methods of soil conservation, i.e. those practices which helped but were not specifically carried out for that purpose. There were some trees left, some mixing of tree and ground crops and other inter-cropping, leaving a cultivated field in large clods rather than a more erodible tilth and refilling yam planting holes from below. These could be used and developed as a basis for more conscious effort in soil conservation. In a survey of explicit attitudes there were the following responses:

1. The soil is strong (fat) or weak, the strongest after fallow.
2. The strength is from juices from decayed grass (not seen in relation to the mineral soil).
3. No mineral soil loss perceived – muddy water dismissed (always occurred) or called plant juices.
4. Some belief in terraces acting to slow down the losses of juice.
5. Anti-government stance – if the government says contour then you don't.
6. Bananas should be clean weeded to prevent root competition.
7. Some could recite soil conservation attitudes without actually doing anything.
8. Only a few actually consciously adopted soil conservation practices such as tree crops and contouring/ditching.

Farm size emerged as an important influence, with the larger farms tending to be owned rather than rented, more trees, more grass and livestock, higher status and income, with any soil conservation projects being focused on the larger farms.

The enablement involved a combination of cash assistance, land tenure and the involvement of people who were already of some standing in the community in "show farms" where not only the techniques but, more importantly, the material benefits were visibly obvious. Markets and social pride were encouraged, especially maintaining the central feature of cash crops, but ones which were not so liable to lead to erosion. Awareness spread and the simpler conservation techniques became evident in the landscape. The elements of success here were ones of sociological sensitivity, working through the existing community structure and the preservation of culturally and economically important activities like going to the market (Plates 8.2 and 8.3).

Plate 8.2 Jamaica, Yallahs Valley. Eroded soil.

Plate 8.3 Jamaica market.

Again, the dialogue between indigenous knowledge and western science is of interest here (Blaikie *et al.*, 1997). The lessons from Blaut and Floyd is that neither indigenous knowledge nor Western science had the answers *per se*. The indigenous knowledge is seen to be inadequate to achieve soil conservation as betrayed by the survey of attitudes. Western knowledge was seen not to be appropriate without calibration for the local community. It is clear that neither is automatically "right" or "wrong". In this context, Lado (1986) in a study of agricultural and environmental knowledge of peasant farmers in southern Sudan wrote that "the view that local knowledge is of little value to planners because it is primitive, partial or biased, is increasingly coming under attack" and continued that traditional and trusted inter-cropping were viewed by agricultural extension officers as inefficient and wasteful, however, the study asserts that this is both ecologically and economically the most efficient technique. Crop diversification mitigates against the failure of one of the crops, the high ground cover reduces the need for weeding and it helps to conserve soil fertility. In fact the need for food security and labour shortage for the task of weeding was why farmers adopted an inter-cropping strategy. This does not mean to say that their agriculture could not be improved, rather that it was an appropriate procedure to adopt in the context in which the farmers found themselves – food risk and labour shortage. Nor does this mean that the indigenous knowledge was "right" but more that the trusted technique was *appropriate*, and that changed circumstances might make other developments more appropriate. Saying that indigenous techniques are somehow right is a form of sentiment which is just as bad as the arrogant assumption that Western science and its techniques are right.

Peasant resilience, adaptation to economic and environmental conditions and underlying attitudes are explored in the book *Pig Earth* by Berger (1979). The author starts with the assertion that peasant life is "committed completely to survival" and quotes Shanin "The peasantry consists of small agricultural producers who with the help of simple equipment and the labour of their families produce mainly for their own consumption and for the fulfilment of obligations to the holders of political and economic power". He goes on to give revealing insights into adoption of innovation. "He may admire knowledge but he never supposes that the advance of knowledge reduces the extent of the unknown." "They envisage the future, to which they are forced to pledge their actions, as a series of ambushes." "The path is a tradition handed down by instructions, example and commentary." "When a peasant resists the introduction of a new technique or method of working it is not because he cannot see its possible advantages – his conservatism is neither blind nor lazy – but because he believes that these advantages cannot, by the nature of things, be guaranteed and that, should they fail, he will then be cut off alone and isolated from the route of survival." "Those working with peasants should take this into account. A peasant's ingenuity makes him open to change, his imagination demands continuity." In fact "conservatism represents a depository of meaning preserved from lives and generations threatened by continual and inexorable change". In other words, experience stands the peasant in good stead in a changing world, making him open to adopt new techniques but only through learning (how and) that they work in practice, not through advice from others. This insight means that what will work is a demonstration that the route to survival is not only not cut off through innovation but that survival is actually enhanced by it. *Indigenous knowledge is thus neither to be deified, neither is Western knowledge to be dismissed.* The criteria by which you judge the dialogue between them is one of demonstrable benefits to survival and livelihood. Thus Lado (1986) sensibly suggests that you should:

1. establish a peasant farmer's knowledge;
2. establish how farmers evaluate their soils and their agricultural potential; and
3. find out what farmers already use to maintain their soil fertility.

Then using this appreciation as a basis, build on it through demonstration, the criteria being ones of resilience and capacity building, which enhance their lives and their ability to cope with change. This however, is neither to characterize tradition as "good" or "bad", it is the dialogue between tradition and new knowledge or awareness that is important. Boserup (1989) in *Women's Role in Economic Development* for example, talks of the possible "tyranny of tradition", with change as a liberating experience. Traditional methods should neither be viewed nostalgically with sentimental attachment, neither is change necessarily a threat. What we learn from Berger is that ingenuity and the repository of meaning from the past can live hand-in-hand – just because things are different it does not mean that they are mutually exclusive.

Further reading

> ** Key reading
> * Important reading
> Recommended reading

Arnold, J.E.M. and Dewees, P.A. (eds) (1997) *Farms, Trees and Farmers: Responses to Agricultural Intensification*. Earthscan.

Banuri, T. and Marglin, A. (1993) *Who will Save the Forests?* Zed Books.

** Blume, H.P., Egere, H., Fleischhauer, E., Hebel, A., Reij, C. and Steiner, K.G. (eds) (1998) *Towards Sustainable Land Use: Furthering co-operation between people and institutions.* Advances in Geoecology 31, International Society of Soil Science.

Reij, C. (1996) *Sustaining the Soil: Indigenous soil and water conservation in Africa*. Earthscan.

* Röling, N.G. and Wagemakers, M.A.E. (1998) *Facilitating Sustainable Agriculture: Participatory learning and adaptive management in times of environmental uncertainty.* Cambridge University Press.

Young, A. (1989) *Agroforestry for Soil Conservation*. CAB International.

* Young, A. 1(998) *Land Resources: Now and for the future*. Cambridge University Press.

Part IV

Conclusions

Chapter 9

Conclusions: environmental change and ecosystem management – attitudes and values

Summary of key points:

1. Contemporary ecological and sociological theory both appear to be talking about the same thing: uncertainty, and capacity building to cope with it.
2. Plurality would appear to allow for a wide range of meanings and also a wide range of responses to unpredictable situations.
3. Approaches cannot be overly prescriptive but should be context-specific, stressing adaptability and a multiplicity of values rather than a tyranny of singular meaning.
4. Nature can be regarded as "what is autonomous to us", with ecosystems as a "contingent state of conjunctions" showing dependency but not necessity.
5. People and other living things have the potential to create new types of responses; development of beliefs and meanings is seen as the way forward in achieving personal and institutional capacity for adaptability.
6. Ideas are enabled into an action through experiences which demonstrate tangible benefits.

This is a surf-riding phase in British Life. The wave of relativism – the obsessive avoidance of judgements of quality, or moral judgement – has risen higher than ever before (as in all prosperous societies). [Education should be a priority] an education not bemused by vocationalism to the detriment of the development of the human critical spirit. [The important thing is] the voting citizen armed with the power of choosing an alternative ... Out of that, a true sense of community, diverse but not divided, might emerge. That'll be the day
Richard Hoggart (1995) *The Way we Live Now*

... green development programmes must start from the needs, understanding and aspirations of individual people, and must work to build and enhance their capacity to help themselves.
Adams (1990) *Future Nature*

Meaning in life is lost by striving after status and future glory; it is gained and realised by action towards a common ideal, in serving the whole according to our physical, mental, educational and revelationary (understanding) capacity.
Mollison and Holmgren (1978) *Permaculture*

I referred to the inertia that results when rules replace thinking, meetings replace doing and process replaces responsibility. In today's fast-moving environment, organisations must learn to become small entities within larger ones, combining the resources and economies of scale of a large corporation with the speed and agility of a start-up company. Both have the same mass but the group of small entities is infinitely faster and easier to steer ... Hire and reward the right people; instil in them the core values of the organisation; give them a clear goal, accountability, and responsibility; and then get out of their way. They can't be energised if they aren't given the leeway to make decisions and act without getting a dozen approvals first.

Plamondon, Chapter 28 in Hesselbein *et al.* (1996) *The Leader of the Future:*
New visions, strategies and practices for the next era

Unless we chose to decentralize and to use applied science ... as a means to producing a race of free individuals, we have only two alternatives to choose from: a militarised totalitarianism ... or, from the social chaos resulting from rapid technical progress ... under the need for efficiency and stability, (we must go) into the welfare-tyranny of Utopia.

Aldous Huxley (1932) *Brave New World*

History teaches that men as a rule do not break with specialised processes which apparently served them well, without the help of disaster ... which ... free(s) the human spirit from the drowning embrace of exclusive conditions.

Laurens van der Post (1978) *Jung and the Story of Our Time*

Individual organisms facing an uncertain future cannot afford consistency.

Drury (1998) *Chance and Change: Ecology for conservationists*

9.1 Thought and action

If we agree that facing the challenges of the future involves both physical and social science, what is interesting is to note the convergence of social science and ecosystem science. In *Contested Knowledge: Social theory in the post-modern era* (Seidman, 1998) the author concludes that what is needed for post-modern sociology are not concepts of system, organism and mechanisms but "approaches to social change that address discontinuous, unpredictable and contradictory trends". This mirrors almost completely Pahl-Wostl's (1995) conclusion that it is appropriate to move away from mechanically viewed ecosystems or seeing vegetation as an organism:

> Instead of trying to become ever more predictive, to squeeze ecological knowledge into a frame that might not be appropriate, it could be the task of ecologists to develop a new type of knowledge to assist in establishing means of how to deal with uncertainties. The urgent need for a change in attitudes has become more evident in reflecting on ... the possible consequences of uncertainty related to the evaluation of the consequences of climatic change on ecosystems.

This is also echoed by Bazzaz (1996) in *Plants in Changing Environments* where, in Chapter 10 (Coping with a variable environment), flexibility of response is also stressed.

Pahl-Wostl then makes a statement which I feel applies equally to the social sciences as to ecosystem sciences: "*Uncertainties should not always be looked upon as being something inherently negative.*" There is an uncertainty that can be viewed in a positive light: "namely the uncertainty associated with the fact of evolution and innovative action. *Living systems obviously have the potential to create new types of responses to unprecedented situations"* (emphasis added). Surely this applies to people as much to plants, animals and other organisms? Houghton (1997, p. 143) certainly feels that it does, stressing the importance of human creativity and the ability to develop further ideas: *"A proper balance between humans and the environment must leave room for humans to exercise their creative skills"*. Figure 9.1 charts the shift in ideas over time (b)–(b) from a more certain positivist approach (a)–(a) to the current era of uncertainty and enablement (c)–(c).

Capacity building and enabling a flexible response must then surely apply equally to social systems and ecosystems. While Berkes and Folke ((1998) *Linking Social and Ecological Systems: Management practices and social mechanisms for building resilience*) see this in terms of resilience, it should be emphasized that this is not just the ability to withstand change, that is to remain exactly the same in the face of changes, it involves *the capacity to respond with flexibility to changed situations while maintaining what is important*. In social systems, this may mean,

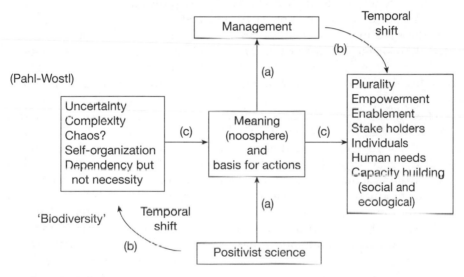

Figure 9.1 Shifts in the relationshiop between ecosystem science and environmental management.

for example, ensuring that a livelihood is economically viable but changing the nature of the work undertaken. In ecosystems it again means enabling a response to change, for example a change in distributions or species composition while maintaining the essential functions of productivity, decomposition and recycling. There may well be elements of inertia, however, though these should not be confused with heritage. Human beings display an attachment to past investments and may be reluctant to relinquish accrued emotional capital; this may display itself as nostalgia or heritage conservation but it is important to remember the force of memory and habit in human nature. People cling on to what has been tried and tested even if changed conditions make it no longer relevant. That is why demonstration of benefit and experience are critical in attitude shifts. In ecology Milchunas and Lauenroth (1995) also talk of inertia, with a tendency to occupy sites which may no longer be favourable. This may manifest itself as a protection of ecosystem heritage which might be seen as important, but this is only possible with considerable effort if the conditions for maintaining that heritage no longer exist. What is more important is the protection of function rather than necessarily the protection of preferred states, though biodiversity is both a preferred state and also functionally important in that there are more potential species to perform the fundamental functions. Here the dialogue between preferred states and functions or processes is important. Similarly, in social situations we have preferred states – well-being, and so on – but the difference is that we may have to change the processes by which we achieve these.

Huxley (1937) in a discussion of ideals and realization in society wrote that "We know what sort of society we should like to be a member of and what sort of men and women we should like to be. But when it comes to decide how to reach the goal, the Babel of conflicting opinions breaks out". He does then admit that the actual goals also vary with each person – or society – which again leads us to a need for enablement where multiple goals can be achieved. But even if we can agree on goals, how are they to be achieved? Goldblatt (1996) in a discussion of social theory and the environment concludes that "democracy may be a necessary condition of making the case for environmental sustainability but it does not guarantee that its arguments will be accepted". The author, however, seems to reach the conclusion that social theory (Marx, Durkheim, etc.) is inadequate. "How to live individually and collectively – how to persuade the selfish, the powerful and the uninterested?" The last lines of the book are that "Neither classical socialism nor contemporary social theory have provided sufficient intellectual or moral resources to answer. We shall have to equip ourselves".

Equipping ourselves means looking at two things: thought and action. It is about finding and facilitating both the attitudes and a set of practices which enhance society and the environment in the capacity to be resilient which is more than simply resisting change but being able to flourish in the face of change. We live in a world of ideas and concepts and a world of (consequent) actions which are codified as a set of practices; Lantermann and Schmitz (1994) in the context of global environmental change writes of the importance of *psychic resources*. The concepts attain meaning and significance through constructs which result from the experience of the interaction between ideas and the consequences of putting them into action. In terms of the role of experience, Kearns (1998) writes how values and experience interact, terming experience "facts". Values prioritize or relegate facts and the feedback is that facts embarrass or endorse values (Figure 9.2).

The basic questions are thus:

- What do we think?
- What do we do?

The former involves knowledge, beliefs, values, constructs, perceptions and interpretations together with experience and memory. The latter is based on a reflexive interaction between the former and the world around us. Goldblatt (1996) writes of the interactions between interests and ideals with social structure and culture in that:

- interests can shape moral claims; and
- growth of knowledge and moral argument can lead to enlightened re-definition of interests.

The author not unnaturally sees the solutions to environmental issues as involving structural causes and political mobilization:

Classical social theory has the capacity to situate political mobilization in the context of:

- broader shifts and structures in modern society; and
- providing an understanding of the interplay of interests, ideals and cognitive understanding of the world.

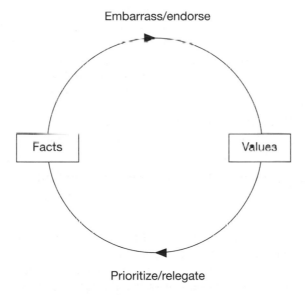

Figure 9.2 The virtual circle of facts and values (from Kearns, 1998).

On page 202 the author says that the success of the green movement rests on the "peculiar forces of better argument" but as I see it, it is claiming the high moral ground which leads to talk about ideals and fails to engage with most people's interests. Indeed, Latour (1998) now writes "Will political ecology pass away?" That is why it is important not just to exhort about recycling but to demonstrate its benefits – and why the work covered in Blume *et al.*'s (1998) volume is important in that it involved demonstrating the benefits of recycling, composting and organic matter use in agriculture in a participatory context where all could see the benefits and realize that *these sets of practices were in their interest.* Thus *the idea is enabled into an action through experience.*

Interestingly, Amanor (1994) writing on adapting changing environments lists five important resources in a West African study of farmers' responses to land degradation; these resources having the potential to assist with adaptability, the production of stable but dynamic economies and economic diversification in the face of changing markets. These resources are:

1. Fallow land (potential further use).
2. Timber products (construction of a variety of products).
3. Fuelwood (for power).
4. Housekeeping provisions – livestock, crops, fibres, medicines, food and beverages, spices, containers, fruits and wild foods.
5. *Human thought* (my italics).

Managed, domesticated and built ecosystems are a result of the interactions of thought and action with the world about us where thought and action have dominated, and increasingly so from managed through domesticated to built. They all produce cultural landscapes. In domesticated agricultural landscapes every piece of land cover is either deliberately created or deliberately allowed. The landscape is the result of maybe many thousands of decisions over time to alter or not to alter. Wilderness systems are where thought and action have led to situations where thought and action do not dominate the world about us, in other words it is the decision not to alter which dominates. Every word we thus use in environmental management is loaded by the link between concept and action: "intervention", "protect", "restore" and so on relate to concepts of nature and human value. "Enable" and "facilitate" should perhaps now be more prominent in attitudes, thought and action involving wilderness, managed, domesticated and built ecosystems and their associated societies.

9.2 The search for meaning: from hot to beautiful and a plurality of meanings

Constructs are a provisional way of dealing with a new reality. If you do not have to test them, or if there is no test of them, they remain as that reality for the individual making that construct. They, of course, can then vary with the individual and hence a debate about contested realities emerges. If something looks hot, we can test that notion of reality through our senses and among a number of people there will be rapid agreement that an object is hot. When we move into areas which are less verifiable, such as landscape, ecosystems and beauty we can construct them in various ways and come up with an number of meanings. Figure 9.3 charts a version of the possible relationships between nature, society and meanings. Deciding which of these is "right" is probably a fruitless endeavour. As Owens (1994) says: "Whilst recognizing that held and assigned values vary between individuals … it becomes necessary to identify what is right and what is good … exactly how this should be achieved …[in] … a participatory planning process, remains one of the least well defined areas of current theory and practice". As we discuss further below, "good" and "right" must surely be seen in terms of outcome and the judgement criteria seen in terms of environmental quality and human interest; the endeavour then is one of devising precise criteria in terms of specific goals (Röling and Wagemakers (1998); see p. 251). However, I might also argue that deciding what is right, while difficult, may not even be necessary. Indeed, rather than framing this as a side-stepping of some kind of moral imperative, it could be seen as important to be able to allow a number of meanings to coexist rather than arguing that what is important is to decide what is right, in other words *a plurality of meanings* is important.

Writing on landscape interpretation, Porteous (1996, 48–49) quotes Meinig's (1979) "beholding eye" concept and cites at least ten modes of viewing the same scene:

1. Nature – humankind's feeble works are overwhelmed by the ageless power of natural forces.

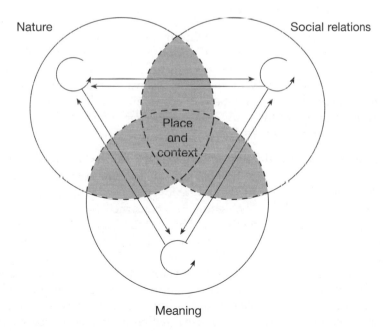

Figure 9.3 The realm of meaning (from Sack, 1990).

2. Artefact – people conquer nature and create humanized landscapes.
3. Habitat – humankind and nature work in harmony to produce an essentially humanized but not dominated earth.
4. System – a dynamic equilibrium of interacting forces (whether physiographic or economic).
5. Problem – applied science which appreciates the system but observes the scene as a problem which can be resolved by application of technical expertise.
6. Resource – the environment as a resource, as a source of wealth. Everything is commodified.
7. Cultural ideology – the landscape reflects the prevailing cultural ideology – the landscape as philosophy translated into tangibility, reflecting dominant value systems (e.g. wealth production, freedom, power, progress).
8. History – a palimpsest of what has gone before; landscape as process and accumulation.
9. Place – a unique entity, a repository of affections of both inhabitants and visitors.
10. Artistic quality of aesthetics, involving both the "pure" (colour, line, mass, texture and balance) and the "less pure" in terms of a landscape of meaning and symbols.

These may be grouped into the more mechanical/utilitarian (artefact, system, problem, resource) and the more abstract (nature, habitat, ideology, history, place, aesthetic).

Porteous also sees Thayer's (1976) notions of visual ecology useful, involving a number of levels of significance:

1. Presentational aesthetic surface – light, line, form, mass, colour, texture.
2. Associative – something with function or use, for example a tree gives shade.
3. Emotional or affective – a tree may calm, affright or interest us.
4. Symbolic – when we measure something against a value system it may symbolize, for example trees as symbolic of economic wealth or of age-old forests.
5. Behavioural or activating – *we chop it down for timber and paper or we sit under the tree and admire its beauty.*

If we are planning for the future, rather than a fruitless discussion of which of these is "right" (where any proposition can be contested because it is unverifiable, unlike "hot"), is not *the coexistence of a plurality of meanings* important? Thus, might not all ten "beholding eye" constructs be tenable in one place to enable people to see what they want? Think of the "serried ranks" of a forestry commission conifer plantation – this really only is driven by, and allows only one construct: that of resource. It is then interesting that, following public opinion, planting of other trees, even flowering cherries, round such uniform resource blocks became adopted in later plantings so at least it might fulfil some other constructs such as aesthetics and cultural (though it is interesting that there have been protests about the normal felling of coniferous forests as they are now seen as part of the landscape, much as there were protests about the original planting). It is, however, clear that plurality does not necessarily resolve conflicts, indeed it may encourage them, as the visual ecology examples show – we may either cut it down or we may admire it. There thus has to be a *dominant meaning* but this does not preclude other meanings. *A productive forest can be habitat, system, cultural, place and have historical significance – it does not just have to be resource.* Tyranny is where only one meaning is tolerated. This does not mean that having only one meaning is bad, however, there may be, and often should be, *a dominant meaning but this should be arrived at normatively, not coercively.*

Porteous (1996, p. 77) writes of an almost schizophrenic attitude to forests quoting a person perched on a mountain crag in a "frontier" area "ambivalently looking east towards 'fruitful fields' emerging from '*dreary forests*', yet also looking westwards towards the *untamed forest* where the '*sublime in nature captivates while it awes and charms while it elevates and expands the soul*' ". However, he has already allowed that the pioneers saw the wilder forests as dismal places (see Chapter 5, p. 110). Forests are thus allowed a plurality of meaning. Firstly, in the context of fruitful farmland, they are out of context because the area is "meant to be" tamed, and are constructed as dreary. Second, the wilderness is constructed as both sublime and dismal. Part of the latter dichotomy is related to "involvement" – the forests are sublime provided you don't have to make a living there. The uncommitted can afford the sublime construct. Bryson (Chapter 5, p. 111) has already been noted as sneering at Thoreau for his romanticized views of nature, that is while he was still safe and near civilization, but once Bryson committed to the wilderness forest, they became fearful places. We might thus distinguish between an essentially *touristic* construct and a *committed* construct of those who live there. This must surely be why symbolic and behavioural levels of significance diverge? Visitors can afford not to have utilitarian constructs. Tourists seek authenticity (May, 1996) but also safety and a decent hotel to return to. Authenticity and a sense of the sublime are thus essentially tourist concepts rather than constructs of the locationally committed and the indigenous.

Thus if man–nature relationships are reflexive (Lowenthal, 1990) that is, referring back to the subjects or relative to the beholder, then coexistence of plurality of meanings of sites and areas and involvements and commitments to sites and areas are seen here as key factors. How, then, do we choose, between the tourist and the committed, the sublime and the utilitarian? Must we choose a dominant meaning? If the latter, then we might go for employment and economic growth; if the former then we might go for forests, flowers and quaint cottage industries – the classic conflict. Can we allow for a plurality of meanings at one site? The answer must be both yes and no, depending on the site. Firstly, we can rely on zonation – some areas set-aside for one dominant meaning and some for another – a spatial resolution of the conflict; second we could be guided by the historical significance of the site – if development destroys something which cannot be replaced, the touristic meaning could dominate over the utilitarian, spatially committed one in order to preserve historical meaning. Thirdly, we have argued that we should facilitate change while retaining what we value. In ecology, it has been argued in the past that diversity promotes stability (if one component is damaged, something else can take its place) but this is essentially an argument about maintaining (preferred) states rather than processes. I would argue slightly differently, diversity enables change. It has the same core – that if circumstances change, something can come to the fore, perhaps something triggered as a "driver" (Holling *et al.*, 1996) which was hitherto not very important. We do not know what will be important in the future, thus diversity should be maintained so that there is greater potential for a wider range of responses. This applies to ecosystems, species, landscapes, meanings and societies. Diversity is thus to be celebrated. I seem to have taken a long time to say "you shouldn't put all your eggs in one basket".

9.3 So: how should we view nature? What do we do for the best – how should we act – what are we trying to achieve and what should we be guided by? Concepts and practices which enable and facilitate ecosystems and societies

First, the answers to these questions lie in an endless search, not in finding prescriptions. Second, if I am to distil any wisdom at all from all I have read and reported herein it is that we should try to believe in something, flawed though it may be. I once asked a group of students how they felt about the future – they were overwhelmingly pessimistic. One of the problems with post-modernism (see Bassett (1996) and also Zimmerman (1994) *Contesting Earth's Future: Radical ecology and postmodernity*) and with deconstruction, which the students had learnt with great alacrity (because it seemed so superior and academic to dissect everything?) is that it can soon turn to destruction – yes, everything is positioned and relative to the beholder, but that is a starting point not a conclusion. As Jeffreys (1962) says in *Personal Values in the Modern World*, cynicism can become a substitute for faith. The point of analysis is, however, not one of destroying everything but to examine underlying assumptions and work out what you actually do, then, believe in. Darier (1999b, 225–227) writes of an "aesthetic of existence" from Foucault and Nietzsche which is a way in which individuals can give meaning to their lives "after the death of external referential points such as 'God' or 'Truth' … we need art … to enable us to carry on being moral in the fate of our recognition of the terror and absurdity of existence". This aesthetic relates to self-examination of morals, thought and behaviour. Harvey (1996) made a key point in this context as: "The task of critical analysis is … to find a more plausible and adequate basis for foundational beliefs that make interpretation and political action meaningful, creative and possible". It is therefore refreshing to read of *useful* research in man–environment themes, for example, in Pacione (1999) in "Applied geography: in pursuit of useful knowledge". There is also a much deeper and useful discussion by Bishop (1992) on "Rhetoric, memory and power: depth psychology and postmodern geography". Bishop indeed relates a "bottomless" post-modern world, with

endless relativizing, but also talks of the benefits of the insight it brings especially, and ultimately of the plurality of imaginative worlds. What one has to
believe in is human ingenuity, the power of thought and of ideas and indeed
in humanity itself – a humanity capable of simultaneously looking after itself
and other life on earth. This, as I have said in this book, is about enabling
capacities and diversities, from a plurality of meanings, to the diversity of
social groups and the diversity of animals and plants, in a non-utopian but
realistic way. Enabling a plurality of meanings also involves enabling a plurality
of practices.

In very concrete terms, a plurality of meanings can include a plurality of
products and activities, which is already a well-developed theme in forestry as
shown by a special volume of *Unasylva* journal (including Anderson *et al.*
(1998) in "Accommodating conflicting interests in forestry – concepts emerging from pluralism" and Vira *et al.* (1998) in "Institutional pluralism in forestry:
considerations of analytical and operational tools"). Here it is seen as important
to have "processes which recognize and involve the multiplicity of ideologies,
interests, objectives and knowledge of the individuals and organizations that
have a stake" in forest management; Jeffrey and Sundar (1998) even write of a
"new moral forestry". Stakeholders are regarded as "all those who affect and/or
are affected by, the policies, decisions, and actions of the system; they can be
individuals, communities, social groups or institutions of any size, aggregation
or level in society. The term thus includes policy makers, planners and
administrators in government and other organizations, as well as commercial
and subsistence user groups" (as quoted by Vira *et al.* from Grimble *et al.*
(1995)). However, as Vira *et al.* point out, this potentially includes the whole
population and they list four principle sets of stakeholders in the context of
forest management:

1. Individuals (private households).
2. Communities and other collective groups.
3. Corporate bodies (commercial enterprises, as well as non-governmental
   organizations).
4. The government.

Establishing common interests and the specification of each partner's role is
then seen as important. O'Riordan (1997) writes of valuation as revelation and
reconciliation – creating trusting and legitimizing procedures of stakeholder
negotiation and mediation.

A parallel approach is also adopted by Röling and Wagemakers (1998)
in: *Facilitating Sustainable Agriculture: Participatory learning and adaptive
management in times of environmental uncertainty.* They use the acronym AKIS
(Agricultural Knowledge and Information Systems) to sum up their approach:

1. Empirically to find linkages – to discover how social actors such as farmers,
   advisors, scientists, credit banks and seed suppliers are linked together in
   sharing, creation, adaptation, storage and application of knowledge.
2. Normatively to design ideal links and flows as a mental construct.

3. Analytically to guide interventions to ensure that the actors do interact in ways that give rise to the desired emergent properties. The goals may be derived widely (agricultural production/sustainability) or narrowly focused (e.g. production of a specified amount of milk from a cow).

The role of supportive institutions is seen as important and some of the key elements include:

1. Granting farmers appropriate property rights.
2. Promoting farmer to farmer exchanges.
3. Directing subsidies and grants to sustainable practices.
4. Linking support payments to resource conservation.
5. Providing information for consumers and the public.
6. Encouraging the formation of local groups.
7. Assigning local responsibilities.
8. Encouraging the formal adoption of participatory processes.
9. Developing capacity in planning for conflict resolution and mediation.

These aspects are also touched on by Anderson *et al.* (1998) in that pluralism can give rise to conflict if not managed carefully. However, as the authors point out, it is not a matter of seeing who is "right" with, perhaps insiders or "locals" being "right" rather than outsiders, but in establishing dialogues between all the players. This admits that there are *several alternative management plans which are consistent with the available scientific evidence* (and thus that science does not have the answer – it is how it is interpreted which matters). It looks at competing claims and values and aims at expanding the range of possibilities. It is actually a question of competing notions, ideas and constructs – but admits that conflicts are inevitable and cannot often be resolved – it tries to allow for and accommodate them. This side-steps the need for consensus. The task is therefore to work with multiple perspectives and possibilities, and not either to shirk from or acquiesce to them. Vira *et al.* (1998) sees this working operationally in terms of the four Rs:

1. Rights to land and forest resources – rights to use the resources (independently of any discussion about the ownership of land).
2. Responsibilities – this is not just decentralized local responsibility, but overall state responsibilities which set the rules as envisaged by Agarwal and Narain (1990) (see Figures 9.4 and 9.5).
3. Returns from the resource – values which motivate the stakeholders.
4. Relationships – especially power relationships, which might need re-negotiation.

It is the question of value and values where contention may arise. In Chapter 5 we mentioned the drive to value ecosystems in monetary terms (Costanza *et al.*, 1997) and also went on to discuss spiritual values in more depth. In terms of the latter, it is clear that many constructs can be made about the same place. We have already idealized that management might attempt to maximize those meanings allowed, but in resource management Briggs *et al.* (1999) have

Components of a village ecosystem management and improvement plan

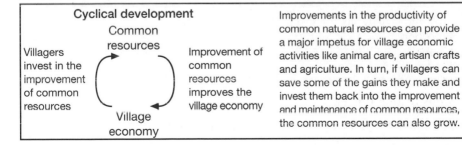

Natural Resource Base

Common resources

Grazing land	Forest land	Ponds and tanks

Private resources

Crop lands	Livestock	Private trees	Wells

Basic Needs

Food	Fuel	Fodder	Manure	Building materials

Artisanal raw materials	Herbs	Drinking water	Irrigation water

Social Structure

Large land holders	Small and marginal land holders	Landless households	Male–female relations

Common property resources can be

Government Property Resources

or

Community Property Resources

But people will react to them differently

They will care for the latter but not the former.

Two steps for rational land use

1 All genetically rich and ecologically sensitive areas should be fully protected with an undisturbed natural forest cover

2 All degraded common lands should be brought under the control of village communities (without necessarily transferring ownership) to protect as common lands and grow grass and trees

Cyclical development

Common resources

Villagers invest in the improvement of common resources

Improvement of common resources improves the village economy

Village economy

Improvements in the productivity of common natural resources can provide a major impetus for village economic activities like animal care, artisan crafts and agriculture. In turn, if villagers can save some of the gains they make and invest them back into the improvement and maintenance of common resources, the common resources can also grow.

Figure 9.4 Green villages (1) (from Agarwal and Narain, 1990).

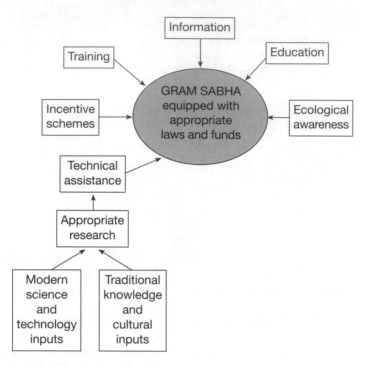

Figure 9.5 Green villages (2) (from Agarwal and Narain, 1990).

usefully written about the value of vegetation in "Indigenous knowledge and vegetation use among the Bedouin". Here, species are ranked in two sorts of preference: direct utilitarian (which may be easily expressed in monetary terms) and less tangible but nonetheless important attributes.

Vegetation, such as *Acacia* and *Tamarix*, are first ranked in terms of their direct value for:

1. Grazing, (separately for sheep, camel and goat).
2. Firewood.
3. Charcoal.
4. Medicine.
5. Shade.
6. Construction timber.

Second, they are ranked for "operational/handling values":

1. Resisting drought.
2. Growing on poor soil.
3. No scorpion/snake attraction.

4. Ease of cutting.
5. Withstanding grazing.

Management may thus be directed to species preference in terms of these values with trade-offs relating to maximizing the range of benefits in different places and according to different needs.

Agarwal and Narain (1990) see the key aspects as:

1. Self reliance.
2. Equity.
3. Concern for the environment (which to me seems to be in their self-interest).

Opie (1998) showed how the Great Plains agricultural policy was nested in a variety of mental constructs. As Colten (1998) comments on the paper "Rooted in popular perception and deep-set values, he finds a powerful force that shaped the agriculture and the successive policy formulations". This is deeply interesting, showing the power of ideas, and the managed landscape as an expression of them. Can we then *change* our ideas deliberatively along the lines suggested above to achieve the desired emergent properties? Zelinsky (1975) was in no doubt of either the weakness of prescription nor of the power of ideas: "In lieu of prescription, I can state some intuitive beliefs. Firstly *the power of ideas.* They matter tremendously, and are *perhaps stronger than any other force on earth* … Secondly … a fundamental revolution may have already begun in human affairs" (emphasis added) and talks of an "astonishing re-ordering of social arrangements, of the conditions of production, consumption, and exchange of goods and services and feelings".

Berkes and Folke (1998) see the key idea as when "successful knowledge and resource management systems will allow disturbances to enter on a scale which does not disrupt the structure and function of the ecosystem and the services it provides". This still seems, however, to relate to meanings of stability which Orians (1975) saw as involving constancy, persistence, inertia and elasticity. But if we follow Pahl-Wostl (1995) and see ecosystems as continually changing and dynamic, and indeed where disturbances are an essential part of the system, we may ask "what happens if the structure and function of the ecosystem is changed so much that the preferred state is no longer there?" How do we react? Sensibly, Harrison (1987) writes that "Forecasts are always dangerous. Human societies are too complex for the future to be predicted with any degree of confidence: the element of free will confounds all attempts at clairvoyance. *People are not automata determined by their circumstances: they can at any time change course and alter those circumstances*" (emphasis added). In that sense, ecosystem change does not matter, we can either seek to alter the ecosystem ourselves or adapt to a changed state. What matters is the capacity to react.

Capacity building is discussed by Bekoe and Prage (1992) and is broadly defined by them as "the acquisition of skills, knowledge and institutions and the generation of motivation to enable specific ends to be achieved", to which I would add the ability to deal with surprise and unforeseen events. They see education and training – bringing knowledge to people – as central to this endeavour, involving individual, group and institutional self-sustained learning (including the capacities and capabilities of indigenous peoples) together with the development of an infrastructure which makes this possible. Fairhead and Leach (1996, p. 188), in a discussion of flexibility and adaptability also stress the considerable flexibility in species which can substitute for each other for a variety of uses – which gives the capacity to obtain fuelwood, fodder, poles and so on from a number of different sources. The species which are able to flourish might vary, but the essential products remain. Writing of African pastoralism, Warren (1995) sees that semi-arid ecosystems are "in permanent disequilibrium" but that many indigenous pastoral strategies are already adapted to this situation. Scoones (1997) rejects simplistic aggregated assessments of people–resource relationships and asks for uncertainty, complexity and non-linear change to be taken seriously, advocating a more embedded, context-specific, adaptive and learning approach. Also, in terms of land resources, Young (1998) seeks policy guidelines where all sections of the community have access to land resources and participate in decision making about their uses. There are thus already many indications of what social and ecological capacity building might involve.

This clearly involves enabling preferred states of (a) the world about us and (b) our being, together with facilitating the ability to respond to change in innovative ways while attempting to keep those things from the past which we value. The key concept of capacity building (of both ecosystems and societies) is participation (allowing for diversity both within society and nature). The key actions are a set or practices which are demonstrably beneficial to people's interests, both in the short term and in the long term (though if it is only in the latter, it is clearly more difficult) and beneficial in terms of preferred states of the world and of our being. The art is then to also democratize knowledge about the links between ecosystem processes and preferred states so that it becomes seen to be important why we should protect processes. In a discussion of "Strategies to enhance adaptability" in the face of climatic change, Goklany (1995) thought that strategies should:

1. increase the ability to feed, clothe and shelter the world's expanding population, regardless of the agent of change;
2. reduce vulnerability of forests, habitats and biological diversity to demographic and other environmental stresses;
3. be compatible with mitigation measures;
4. be independent of results from more detailed and accurate site-specific impacts assessments which will be unavailable for several years;
5. be implementable today; and
6. have clear benefits now and in the future.

Utopia is of course a chimera – a Cheshire cat. Huxley wrote in 1937 about a "cure" for war but perhaps accepts the inevitability: "Every road towards a better state of society is blocked, sooner or later, by war ... that is the truth, the odious and inescapable truth" and he starts his book *Island* (Huxley, 1962) with a quote from Aristotle: "In framing an ideal we may assume what we wish, but we should avoid impossibilities". We are back to Gandhi's *realistic idealism* discussed by Agarwal (1987) in *Gandhi, Ecology and the Last Person*. Wakeford and Walters (1995) asked whether science could make the world a better place. Now perhaps we can see some realistic way forward. There are to my mind three sets of activities.

First, the task for the social scientists is perhaps to address structural issues – cars pollute, but why so many cars, lax emissions controls and poor public transport (Goldblatt, 1996)? Rowell (1996) puts it more graphically "Find a global economic system that provides work without destroying the world". And as we said at the start, finding a framework for dealing with unpredictability – capacity building.

Second, the task for science is to demonstrate the links between ecosystem processes and preferred states, how preferred states might vary with environmental change and indicate means of protecting those processes and thus the preferred states (which may be re-defined in the face of change).

There is, however, a third endeavour which is a hybrid one (Forsyth, 1996). This is very much focused on the experiential relationship between thought and appropriate action, defining appropriate as provided for human welfare/interest and environmental quality/preferred ecosystem state. This has to be participatory and allow for a range of interests. Writing of the "green backlash" Rowell (1996) discusses how to beat the backlash against environmentalism. Environmental groups are changing direction: "this will mean going back to the grassroots and start organising, campaigning and talking face to face, door to door, street to street, community to community. Because this is the only way to start rebuilding a definite agenda and a positive message". To this I would perhaps add "*listening*", rather than talking/campaigning, in order to gauge the range of views.

In this context it is interesting to read the Oxfam handbook (Pratt and Boyden, 1985) which deals with ideals and actual practices in situations in social contexts. This kind of manual is central to the translation of ideals into realistic contexts as the intentions are clear but the prescriptions can be calibrated for specific contexts. It is much more than the fruitless polarization between "Western science" (seen as good, especially in the past and more now as bad – Banuri and Marglin, 1993) and "indigenous knowledge" (seen as inadequate to tackle issues or as the answer and way forward). Such a polarization is unhelpful as it is the dialogue between the two which is what matters and the distillation of the good parts of each. "Good" is judged by the outcome in terms of human interests and environmental quality, so what matters then is how you measure the success of the operation in terms of these criteria:

1. Ideas and concepts which are held.
2. Translation of them into specific contexts.
3. Proposition of action.
4. Demonstration/experience/participation.
5. Judgement of usefulness/appropriateness.
6. Revisit items 1–3, ideas, concepts and their translation into action.

In this context, and say, in agriculture, land use practice is critical. What you do, and the effects of what you do, is important as well as the important concept that what you do will have wider consequences than the intended target.

So: how should we view nature?

OK, tick one box.

What is nature?

benign ☐
resilient ☐
fragile ☐
vulnerable ☐
linear ☐
dynamic ☐
chaotic ☐
self-defining/correcting (Gaia) ☐
non-linear ☐
all of the above ☐
none of the above ☐

Deciding between these views really isn't the point. We can find evidence for all of them or against all of them if we are so minded. We can only view nature through our conceptual filters and thus there is no independent proof of any of them. We certainly have preferred states – and these include utilitarian criteria of productivity and a diversity of ecosystem goods as well as the more spiritual criteria such as naturalness. O'Riordan and Jordan (1999) classify people with different world views and their attendant values of nature (Figure 9.6). However, whether nature is seen as capricious, tolerant, resilient or vulnerable, there is something yet more fundamental.

This is the relationship between man and nature, much discussed by Evernden (1992), Braun and Castree (1998) and many others as involving the social construction of nature. The relationship is, in essence, the relationship between will and autonomy – internal human will and an autonomous existence externally. *Nature is what is autonomous to us.* Nature is what has an existence independent of our will. We might plant a tree (expression of our will) but we do not grow it (expression of nature's autonomy). Olwig (1996) calls it "The organic world as experienced by the human community". As such nature is valued as something we can sense other than ourselves, giving us a sense of something beyond us. Our wills interact with this sense, through a range from allowing that sense to come into us: "the mind claiming to itself some of the dignity of the things which it contemplates" (p. 115) to allowing our will to exert itself over the external autonomy. Thus, on the one hand we have the wilderness experience, nature rambles and the countryside where we allow that external sense in to us and on the other we have agriculture, forestry and domestication where we allow our will out from us onto that autonomy. Gardening, as an expression of the man–nature relationship, allows for both flows but again with

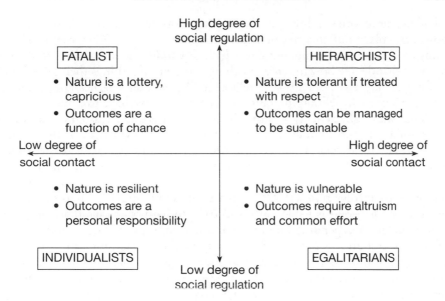

Figure 9.6 Institutions (from O'Riordan and Jordan, 1999).

a spectrum from the wild garden where nature's autonomy is allowed to the neatly clipped lawn and formal border where our will is more dominant.

Beyond that, any construct is cultural and nature is indeed defined by how we see it – by internalized constructs about what is external. As Bishop (1992) says, there are then "images that dwell in the place 'in between' the object and the viewer". Any logic, structure or other concept we have about nature is our image, our imposition. Factors, dependency and so on are all expressions of the human will and the endeavour to arrange something autonomous as if there were rules. We are Hugo's Gilliatt (p. 113) staring into the darkness, underlined by uncertainty. It is not until you have gazed into negation, into utter nothingness, that you have the enlightenment that you have to fill the void with your own beliefs and self-faiths and the knowledge that you can be truly creative. We have peopled the darkness with gods, myths, and explanations. Walk into a wood: it does not know it is an ecosystem, recycling is our concept, structure is our concept – in our search for logic we impose our own limitations – Bishop. "Images do not raise consciousness, they force it to descend". We can always discover pattern and, as we saw in Chapter 3, Pimm (1991) observed that "if you look at random data long enough, all kinds of patterns can be imagined" and "community ecology is so conceptually complex that we rarely have any difficulty in explaining any result we encounter. Theories tell us where to look and when we readily find what we are looking for, we gain confidence in both the theories ... and in the data". This does not mean to say that pattern is there for us to discover, *pattern is a provisional identification not inconsistent with the evidence we can find in relation to our concepts.* "Understanding" means that we can find support for our theories, which is immensely comforting.

Challenging theories is thus disturbing, as are notions of chaos. Chaos does not however replace our theories of structure, chaos is just another way of seeing things and neither structure nor chaos have any independent justification. "It is humans that decide how to represent things, and not the things themselves" and "telling it like it is ... is also ... telling it like we are" (Barnes and Duncan, 1992). So is a self-defining plurality of meanings all we have to offer?

What we might actually venture, in terms of Bertrand Russell's "warranted assertability" (p. 43), rather than in the sense of any "evidence", is that nature and ecosystems are about *conjunction, history and contingency*. Factors coincide in many combinations over time and there is no necessity in their conjunction, just the "thusness" of them which has occurred over history. Contingency, as discussed by Fiedler *et al.* (1997), is that while something is seen as dependent on something else, it is liable but not certain to happen. Thus, there might be many outcomes depending upon the conjunction of prior states, combinations of factors and the turn of events. This is neither chaos nor order nor even self-organization but it is *a contingent state of conjunctions*: a state which has come into existence over time and in relation to the conjunction of events and processes which *show dependency but not necessity*.

What do we do for the best – how should we act – what are we trying to achieve and what should we be guided by?

If we are to be guided by the above it seems reasonable to argue that we should find ways of meshing our will with nature's autonomy. In protecting nature we are protecting its autonomy and we might go on to say that anything which acts to detract from this might be seen as undesirable. This might justify anti-pollution measures and wild life sanctuaries but it is not enough because it also allows plagues of locusts, mosquitoes and malaria to exist. Everywhere we are expressing our preferences and are involved with our sense of meaning.

Operationally, prescriptive "solutions" based on current constructs may not be adequate and ecological and social capacity building should be advocated. This is seen in terms of concepts, institutions, attitudes and values which allow for a plurality of meanings and which can cope with surprise and unforeseen change – and which also facilitates responses to change. This endeavour is actually worthwhile independently and irrespective of any prospect of climatic change and is surely something we should be doing anyway.

Philosophically, the notion of Gaia results in the tenet of robustness in that the earth can look after itself (almost despite whatever we may throw at it?), while the notion of fragility evinces the notion of stewardship. However, the fundamental question posed both by Proctor (1995) and Demeritt (1994) amongst others is "what kind of world do you want to live in?" I have read some writings about utopian solutions to the world which then despair about what to do with greed, stupidity and so on. Such hopeless analyses fail to allow for the range of human responses. Surely self-interest should be legitimated and not just in material terms but also in spiritual terms. In many ecological writings there is the underlying elitist theme of having to make stupid people respect nature and species – we just have to convince them of their worth. Most people do not recognize this at all. The answer is usually seen in terms of education. Huxley (1937) in *Ends and Means* saw this in terms of education for freedom and responsibility. But "what should we tell the children?"

There is a now a wide debate about having "an environmental ethic" or a plurality of approaches (see, for example, Trudgill (1999), Davies (1999), Palmer (1998a) in *Environmental Education for the 21st Century* and Cooper and Palmer (1995) in *Just Environments*). In the first mentioned publication I wrote:

> There is a wide spectrum of human behaviour but there is usually a central tendency of modality. Such a normative situation is usually reinforced through early training, education, media and other inputs. There is some kind of agreed cultural agenda for training, either implicit or explicit in an educational curriculum. Many of the items pertain to the way in which people behave with regard to each other and towards other things they experience. The attitudes and behaviour engendered give the society in which the individual lives its characteristics. Individuals deviating from the norm may be tolerated, ridiculed ignored or sanctioned in some way by other individuals' reactions or more formally through corporate societal actions leading some form of tolerant marginalisation or explicit exclusion. Thus in some societies killing another individual by an individual is not tolerated while in others it may be seen as justifiable in certain cases, and so on. In terms of dress codes, for example, many conform while a wide range of individuality is also tolerated. Where do we stand on environmental ethics? Do we have a standard set of codes to which we expect most people to conform (and impose sanctions when they do not) or is a wide range of views tolerable? In short, do we tell the young what is right and wrong or do we let them make their own minds up through experience? We might readily teach that murder and theft are wrong and we might readily encourage or allow self-expression and individuality in say clothing or art but where should environmental teaching lie in this spectrum?

> I believe that the way forward is to have a centralised environmental ethic which is demonstrably for people's material and spiritual benefit. What we should thus be teaching is not doom and gloom, with admonishments and exhortations which simply turn people off but positive frameworks which teach that it is right to conserve the environment and its organisms. This is not just an economic valuation of environmental and ecological commodities but also a spiritual one. From this central framework, one then creates a range of first hand and vicarious experiences from which people can learn what it might mean for them, given that real identification and meaning comes best through living through an experience.

> What might this mean in practice? First, *the environment is for you* should be the central message – after all people will only think "what's in it for me (both materially and spiritually)?" Second, experience and choice are paramount. Ethics (see Berry (1999) and Proctor (1998b)) develop through being given choices, seeing the possible and probable outcomes and being able to judge the merits of the outcomes (Smith, 1995; Uzzell et al., 1995). This can be achieved in the classroom through videos together with computer games, group activities and discussions. Also very important is fieldwork but again involving discussions about choices for the future as well as experiences of environment and ecology: engendering the ethic of leaving things you have seen and enjoyed for others to enjoy can only be done by seeing a

plant, or a bird at first hand. Fundamental to this is that ecologists can't just expect people to automatically treasure a plant because it is rare – there has to be some meaningful experience which a person can relate to before this ethic is engendered … Making environment and ecology relevant to people's lives is the challenge for ecologists, not adopting the higher ground and admonishing people to behave better.

"Humanity displays two faces, one of drama and the other of indifference" (Antoine de Saint-Exupery, 1939). Without engagement, there is no drama. Without experience, there is no learning: "the crucial components (are) extending personal *experiences*, developing personal *concerns* and promoting personal *actions*" (Palmer, 1998a in *Environmental Education for the 21st Century*.) She lists the important factors in spiritual education (Palmer, 1998b) as involving:

- Personal beliefs and identity.
- A sense of awe, wonder and mystery.
- Experiencing feelings of transcendence.
- Search for meaning and purpose.
- Self-knowledge.
- Relationships – recognizing the worth of other individuals.
- Creativity.
- Feelings and emotions.

While we have to produce food and other material needs, these values systems underpin everything we do. We should try to be creative, not just conservationist (Bell and Evans, 1997). Phifer (1999), reviewing Grumbine's *Ghost Bears: Exploring the Biodiversity Crisis* likewise feels that that author "shows there are multiple ways to value and understand our world, and he calls for management policies that more comprehensively incorporate these diverse perspectives".

To incorporate such a diversity, the steps involved might be:

1. To move from a singular, dominant meaning of "Nature is all", and this might give way to:
2. People are important (rather than espousing an ecological imperative), however
3. It should be realized that "people" is not a homogeneous concept and that while there might be consensus, people will never totally agree, so there will be a range of opinions and perspectives.
4. Therefore plurality is important: this approach specifically allows for a range and diversity of views, but
5. How do you put this into practice? And:
6. While the enablement of diversity can be laudable, we can still ask whether there still has to be a higher ideal? We then look to shared values and end up with the democratic questions: we can, we do, what can we be. Given that we are self-referential and reflexive, we are still left with the question: what do we want?

The burden of being human is that you have to make up your own mind what to do: "to be truly human means a constant state of becoming" (Zelinsky, 1975). The burden of institutions should be to facilitate the thought processes involved. Plurality allows the individual to feel valid. Beyond that, "it is the poet who tells us everything about meaning, the scientist little" (quote from Bishop, 1992) . This is not to demean the scientists, especially if science is seen as it is by Dunbar (1995) as a "continuous process of criticism" which means it is absolutely fundamental. However, it is the poet who gives the meaning. It is the world of meaning that is between the viewer and the object, between the self and an autonomous nature. Education is commonly seen as a panacea, and yes, there is always a need for information, of awareness, especially of the consequences of action, but we live in worlds of meaning and significance. This world of meaning is often termed the "noosphere" (see Samson and Pitt, 1999) who see it as the intersection of science and philosophy where science forms a basis for understanding but falls short of definitive answers. The task is to try to gauge what meanings there are for individuals and groups, cherishing diverse cultural experiences (Di Chiro, 1995) and work with them to build up a respect for the "other" in nature and a respect for the needs of the self, making them both compatible and able to respond to change despite previous investment. The endeavour is self-empowerment with eco-social capacity building. Poetry is a better guide than science. Meaning is the key.

We have sought to understand nature through cause and effect models and manipulated putative causes in order to achieve or maintain particular effects. However, as Pahl-Wostl (1995) observes "cause and effect relationships are more or less non-existent in ecosystems, effects are contingent on state and context". Thus, if states are a matter of conjunction rather than of causes, it may be futile to try to maintain particular states through manipulating putative causes. Therefore, my feeling is that we would be better to attempt to find meaning through conserving particular ecological and sociological processes and mechanisms rather than aiming to achieve particular states. In times of environmental change, this approach would facilitate processes which can cope with change – rather than us trying to shore up states where the conditions which maintain them no longer obtain. The understanding of processes is surely within the realm of science but the maintaining of states is surely cultural and within the realms of preferences and meaning: *ecological process is a matter for science; ecosystem state is a matter for cultural preferences.* There are those who would underscore a critical role for science in environmental management and there are those who, by contrast, deny science a role, seeing it as elitist. Democratizing specialist ecological knowledge is indeed a tall order – but an unnecessary task to face in its potential entirety. Ecological science can and should contribute as publicly and as clearly as possible to the debate about future landscapes and ecosystem states. However, it is only one voice; democratizing the debate about preferred ecosystem states and their relationships with human well-being is the real challenge.

Further reading

```
***  Essential reading
**   Key reading
*    Important reading
     Recommended reading
```

Adams, W.M. (1990) *Green Development: Environment and sustainability in the third world.* Routledge.

*** Berkes, F. and Folke, C. (eds) (1998) *Linking Social and Ecological Systems: Management practices and social mechanisms for building resilience.* Cambridge University Press.

Berry, R.J. (1999) A worldwide ethic for sustainable living. *Ethics, Place and Environment*, 2, 1, 97–107.

* Bishop, P. (1992) Rhetoric, memory and power: depth psychology and postmodern geography. *Environment and Planning D: Society and Space*, 10, 5–22.

** Brown, V., Smith, D.I., Wiseman, R. and Handmer, J. (1995) *Risks and Opportunities: Managing environmental conflict and change.* Earthscan. Especially Chapter 16, Individual learning and personal change, and Chapter 17, Organisational learning and social change, 309–329.

Bruenig, E.F. (1996) *Conservation and Management of Tropical Rainforests: An integrated approach to sustainability.* Commonwealth Agricultural Bureau.

Cooper, D.E. and Palmer, J. (1998) *Spirit of the Environment: Religion, value and environmental concern.* Routledge.

*** Darier, E. (ed.) (1999a) *Discourses of the Environment.* Blackwell. Especially Chapter 11, Darier, E., Foucault against environmental ethics, 217–240.

Evernden, N. (1992) *The Social Construction of Nature.* Johns Hopkins.

* Goldblatt, D. (1996) *Social Theory and the Environment.* Polity Press.

Lowenthal, D. (1993) Awareness of human impacts: changing attitudes and emphases. Chapter 8 in Turner, II, B.L., Clark, W.C., Kates, R.W., Richards, J.F., Mathews, J.T. and Meyer, W.B. (1993) *The Earth as Transformed by Human Action: Global and regional changes in the biosphere over the past 300 years.* Cambridge University Press, 121–135.

Olwig, K. (1996) Nature – mapping the ghostly traces of a concept. Chapter 3 in Earle, C., Mathewson, K. and Kenzer, M.S. *Concepts in Human geography*, Rowan and Littlefield, London, 63–96.

Palmer, J. (1998a) *Environmental Education for the 21ˢᵗ Century.* Routledge.

** Proctor, J.D. (1998b) Geography, paradox and environmental ethics. *Progress in Human Geography*, 22, 2, 234–255.

Rowell, A. (1996) *Green Backlash: Global subversion of the environmental movement.* Routledge.

Sack, R.D. (1993). The realm of meaning: the inadequacy of human–nature theory and the view of mass consumption. Chapter 40 in Turner, II, B.L., Clark, W.C., Kates, R.W., Richards, J.F., Mathews, J.T. and Meyer, W.B. (1993) *The Earth as Transformed by Human Action: Global and regional changes in the biosphere over the past 300 years.* Cambridge University Press, 659–671.

* Samson, P.R. and Pitt, D. (eds) (1999) *The Biosphere and Noosphere Reader: Global environment, society and change.* Routledge.

Trudgill, S.T. (1999) Environmental Education, Ethics and Citizen Conference held at the Royal Geographical Society (with the Institute of British Geographers), 20 May 1998. Introduction, and Postscript. *Ethics, Place and Environment*, 2, 1, 81–82, 87–89.

Zelinsky, W. (1975) The Demigod's dilemma. Annals of the Association of American Geographers, 65, 2, 123–142.

Zimmerman, M.E. (1994) *Contesting Earth's Future: Radical ecology and post-modernity*. University of California Press.

References

Adams, L. W. and Dove, L.E. (1989) *Wildlife Reserves and Corridors in an Urban Environment*. National Institute for Urban Wildlife, Columbia, Maryland, USA.

Adams, J., Maslin, M. and Thomas, E. (1999) Sudden climate transitions during the Quaternary. *Progress in Physical Geography*, 23, 1, 1–36.

Adams, W.M. (1990) *Green Development: Environment and sustainability in the third world*. Routledge.

Adams, W.M. (1996) *Future Nature: A vision for conservation*. Earthscan.

Adams, W.M. (1997) Rationalization and conservation: ecology and the management of nature in the United Kingdom. *Transactions of the Institute of British Geographers*, 22, 277–291.

Adelsohn, U. (1994) From a personal point of view: integrated measures to overcome barriers to minimise harmful fluxes from land to water. *Proceedings of the Third International Stockholm Water Symposium*, 10–14 August 1993. Stockholm Vatten AB/Stockholm Water Company, 73–78.

Aerts, R. (1995) The advantages of being evergreen. *Trends in Ecology and Evolution*, 10, 10, 402–407.

Agarwal, A. (1987) *Gandhi, Ecology and the Last Person*. 13th Gandhi Peace Foundation Lecture, January 30, 1987. Centre for Science and Environment, New Delhi.

Agarwal, A. (1995) Dismantling the divide between indigenous and scientific knowledge. *Development and Change*, 26, 4, 413–439.

Agarwal, A. and Narain, S. (1990) *Towards green villages: a strategy for environmentally sound and participatory development*. New Delhi: Centre for Science and Environment.

Ahlback, A.J. (1995) Mobilizing rural people in Tanzania to tree planting – why and how. *Ambio*, 24, 5, 304–310.

Amacher, G.S., Hyde, W.F. and Kanel, K.R. (1996) Household fuelwood demand and supply in Nepal Tarai and Midhills – choice between cash outlays and labour opportunity. *World Development*, 24, 11, 1725–1736.

Amanor, K. (1994) *The New Frontier: Farmer responses to land degradation, a West African Study*. Zed Books.

Anderson, D.L.T. and Willebrand, J. (1996) *Decadal Climatic Variability: Dynamics and predictability*. NATO ASI Series 1: Global and Environmental Change, Vol. 44.

Anderson, J., Clement, J. and Crowder, L.V. (1998) Accommodating conflicting interests in forestry – concepts emerging from pluralism. *Unasylva*, 49, 3–10.

Angell, W.H. (1994) The wilderness solo – an empowering growth experience for women. *Women and Therapy*, 15, 3–4, 85–99.

Anon. (1999) Deep thinking solves climate change problem. *Chemistry and Industry*, 10, 17 May, 373.

Ashby, E. (1978) *Reconciling Man with the Environment*. Oxford University Press.

Asher, S.J. (1994) Therapeutic considerations of wilderness experiences for incest and rape survivors. *Women and Therapy*, 15, 3–3, 161–174.

Atkinson, A. (1991) *Principles of Political Ecology*. Belhaven.

Avery, D. (1997) The folly of organic farming. *Chemistry and Industry*, 24, 15 December 1997, 1014.

Bak, P. (1997) *How Nature Works: The science of self-organized criticality*. Oxford University Press.

Bandyopadhyay, R. (1996) Global review of land reform – a critical perspective. *Economic and Political Weekly*, 31, 11, 679–691.

Banuri, T. and Marglin. A. (1993) *Who will Save the Forests?* Zed Books.

Barbier, E.B. (1987) The concept of sustainable development. *Environmental Conservation*, 14, 2, 101–110.

Barnes, T. and Duncan, J. (eds) (1992) *Writing Worlds: Discourse, text and metaphor in the representation of landscape*. Routledge.

Barraclough, S.L. and Ghimire, K.B. (1995) *Forests and Livelihoods: The social dynamics of deforestation in developing countries*. UNRISD/Macmillan Press.

Barry, R.G. and Chorley, R.J. (1998) *Atmosphere, Weather and Climate*. Routledge.

Bassett, K. (1996) Postmodernism and the crisis of the intellectual: reflections on reflexivity, universities, and the scientific field. *Environment and Planning D: Society and Space*, 14, 507–527.

Batterbury, S., Forsyth, T. and Thomson, K. (1997) Environmental transformations in developing countries: hybrid research and democratic policy. *The Geographical Journal*, 163, 2, 126–132.

Bazzaz, F.A. (1996) *Plants in Changing Environments: Linking physiological, population and community ecology*. Cambridge University Press.

Beckley, T.M. and Korber, D. (1995) Sociology potential to improve forest management and inform forest policy. *Forestry Chronicle*, 71, 6, 712–719.

Beerling, D.J., Chaloner, W.G., Huntley, B., Pearson, J.A. and Tooley, M.J. (1993) Stomatal density responds to the glacial cycle of environmental change. *Proceedings of the Royal Society of London, B, Biological Sciences*, 251, 1331, 133–138.

Beese, F.O. (1996) Indicators for a concept of multifunctional forest use. *Forstwissenschaftliches Centralblatt*, 115, 2, 65–79 (English summary from BIDS).

Beevoir, K. (1993) *A Tuscan Childhood*. Penguin Books.

Bekoe, D.A. and Prage, L. (1992) Capacity building. Chapter 14 in Brennan, M. (compiler) *An Agenda of Science for Environment and Development into the 21st Century*. International Council of Scentific Unions/Cambridge University Press, 256–263.

Bell, M. and Evans, D.M. (1997) Greening "the heart of England" – redemptive science, citizenship and "symbol of hope for the nation". *Environment and Planning D: Society and Space*, 15, 257–279.

Bell, P.A., Greene, T.C., Fisher, J.D. and Baum, A. (1996) *Environmental Psychology*. Harcourt Brace.

Bendall, R. (1996) Biodiversity: the follow up to Rio. *The Globe*, UK Global Environmental Research, 30, April.

Bennet, A.J. (1994) Soil science and better land use in the tropics. Chapter 20 in Syers, J.K. and Rimmer, D.L. *Soil Science and Sustainable Land Management in the Tropics*. CAB International, 275–282.

Bennett, R.J. and Chorley, R.J. (1978) *Environmental Systems*. Methuen.

Berg, B., McClaugherty, C., Desanto, A.V., Johansson, M.B. and Ekbohm, G. *et al.* (1995) Decomposition of litter and soil organic-matter – can we distinguish a mechanisms for soil organic matter buildup? *Scandinavian Journal of Forest Research*, 10, 2, 108–119.

Berger, J. (1979) *Pig Earth*. Writers and Readers Publishing Cooperative.

Berkes, F. and Folke, C. (eds) 1998) *Linking Social and Ecological Systems: Management practices and social mechanisms for building resilience*. Cambridge University Press.

Berkes, F. (1999) *Sacred Ecology: Traditional ecological knowledge and resource management*. Taylor & Francis.

Berry, R.J. (1999) A worldwide ethic for sustainable living. *Ethics, Place and Environment*, 2, 1, 97–107.

Bhaskar, R. (1975) *A realist theory of science*. Hassocks Harvester Press.

Bird, E.A.R. (1987) Social construction of nature: theoretical approaches to the history of environmental problems. *Environmental Review*, 11, 255–264.

Birkeland, P.W. (1984) *Pedology, Weathering and Geomorphological Research*. Oxford University Press.

Bishop, P. (1992) Rhetoric, memory and power: depth psychology and postmodern geography. *Environment and Planning D: Society and Space*, 10, 5–22.

Blaikie, P. (1995) Changing environment or changing views? A political economy for developing countries. *Geography*, 80, 203–214.

Blaikie, P. and Brookfield, H.C. (eds) (1987) *Land Degradation and Society*. Methuen.

Blaikie, P., Brown, B., Dixon, P., Sillitoe, P. Stocking, M. and Tang, L. (1997) Knowledge in action: local knowledge as a development resource, and barriers to its incorporation in natural resource research and development. *Agricultural Systems*, 65, 2, 217–238.

Blaut, J. M., Blaut, R.P., Harman, N. and Moerman, N. (1959) A study of the cultural determinants of soil erosion and conservation in the Blue Mountains of Jamaica. *Social and Economic Studies*, 8, 403–420.

Blume, H.P., Egere, H., Fleischhauer, E., Hebel, A., Reij, C. and Steiner, K.G. (eds) (1998) *Towards Sustainable Land Use: Furthering co-operation between people and institutions*. Advances in Geoecology 31, International Society of Soil Science.

Bly, R. (1990) *Iron John: A book about men*. Addison-Wesley Publishing Company Inc.

Boehmer-Christiansen, S. (1994) Global climate protection policy – the limits to scientific advice. *Global Environmental Change – Human and Policy Dimensions*, 4, 2, 140–159.

Bolin, B., Doos, B.R., Jager, J. and Warrick, R.A. (1986) *The Greenhouse Effect, Climatic Change and Ecosystems*. Wiley.

Bonnett, A. (1996) The new primitives: identity, landscape and cultural appropriation in the mythopoetic men's movement. *Antipode*, 28, 3, 273–291.

Booth, D.E. (1997) Preserving old growth forest ecosystems: valuation and policy. *Environmental Values*, 6, 31–48.

Bormann, F.H. and Likens, G.E. (1979) *Pattern and Process in a Forested Ecosystem*. Springer-Verlag.

Bornkamm, R., Lee, J.A. and Seaward, M.R.D. (1981) *Urban Ecology*. Blackwell.

Boserup, E. (1989) *Women's Role in Economic Development*. Earthscan.

Bouwman, A.F. (ed.) (1990) *Soils and the Greenhouse Effect*. Wiley.

Bowler, I. (1996) *Agricultural Change in Developed Countries*. Cambridge University Press.

Box, E.O., Crumpacker, D.W. and Hardin, E.D. (1989) Predicted effects of climatic change on distribution of ecologically important native tree and shrub species in Florida. *Climatic Change*, 41, 213–248.

Boyd, W. (1991) *Brazzaville Beach*. Penguin Books.

Brady, N.C. (1995) *The Nature and Properties of Soils*. Macmillan.

Bramwell, A. (1989) *The Literary Ecologist*. Yale University Press.

Braun, B. and Castree N. (eds) (1998) *Remaking Reality: Nature at the millennium*. Routledge.

Brennan, A. (1988) *Thinking about Nature: A investigation of nature, value and ecology*. University of Georgia Press, Athens, Georgia.

Breymeyer, A.I., Hall, D.O., Melillo, J.M. and Argren, G.I. (eds) (1996) *Global Change: Effects on coniferous forests and grasslands*. SCOPE 56. Wiley.

Briggs D.J. and Courtney, F.M. (1985) *Agriculture and Environment*. Longman.

Briggs, D.J. and Smithson, P.A. (1989) *Fundamentals of Physical Geography*. Unwin Hyman.

Briggs, J., Badri, M. and Mekki, A-B (1999) Indigenous knowledge and vegetation use among the Bedouin in the eastern Desert of Egypt. *Applied Geography*, 19, 2, 87–103.

Bronowski, J. and Mazlin, F. (1963) *The Western Intellectual Tradition*. Penguin.

Brookfield, H. and Stocking, M. (1999) Agrodiversity: definition, description and design. *Global Environmental Change*, 9, 77–80.

Brouwer, I.D., Hoorweg, J.C. and VanLiere, M.J. (1997) When households run out of fuel: Responses of rural households to decreasing fuelwood availability, Ntcheu District, Malawi. *World Development*, 25, 2, 255–266.

Bruenig, E.F. (1996) *Conservation and Management of Tropical Rainforests: An integrated approach to sustainability*. Commonwealth Agricultural Bureau.

Bryson, B. (1997) *A Walk in the Woods*. Black Swan.

Buckley, G. (1998) The environmental transformation of an Appalachian valley, 1850–1906. *The Geographical Review*, 88, 2, 175–198.

Bunyard, P. and Morgan-Grenville, F. (1987) *The Green Alternative: A guide to good living*. Methuen.

Burke, M.J.W. and Grime, J.P. (1996) An experimental study of plant community invasibility. *Ecology*, 77, 3, 776–790.

Burroughs, W. (1997) *Does the weather really matter? The social implications of climatic change*. Cambridge University Press.

Burrows, C.J. (1990) *Processes of Vegetation Change*. Unwin Hyman.

Burt, T.P. and Shahgedanova, M. (1998) An historical record of evaporation losses since 1815 calculated using long-term observations from the Radcliffe Meteorological Station, Oxford, England. *Journal of Hydrology*, 205, 101–111.

Button, J. (1989) *How to be Green*. Century.

Byatt, A.S. (1997) *Unruly Times: Wordsworth and Coleridge in their time*. Vintage.

Cameron, L. (1999) Histories of disturbance. *Radical History Review*, 74, 4, 4–24.

Cameron, L. and Forrester, J. (2000) "A nice type of English scientist": Tansley, Freud and a psychoanalytic dream. *History Workshop Journal*, 48, 64–100.

Carlson, T.N. and Bunce, J.A. (1996) Will a doubling of carbon dioxide concentration lead to an increase or decrease in water consumption by crops? *Ecological Modelling*, 88, 1–3, 241–246.

Carpenter, R.A. (1996) Uncertainty in managing ecosystems sustainably. Chapter 4 in Lemons, J. *Scientific uncertainty and environmental problem solving*. Blackwell Science, 118–159.

Carson, R. (1962) *Silent Spring*. Penguin.

Cernea, M. (1988) Alternative social forestry development. Chapter 6 in Ives, J. and Pitt, D.C. (eds) *Deforestation: Social dynamics in watersheds and mountain ecosystems*. Routledge, 159–190.

Cernea, M.M. (ed.) (1991) *Putting People First: Sociological variables in rural development*. Oxford University Press.

Chamberlain, D.E., Wilson, J.D. and Fuller, R.J. (1999) A comparison of bird populations on organic and conventional farm systems in southern Britain. *Biological Conservation*, 88, 3, 307–320.

Chapin, F.S., Rincon, E. and Huante, P. (1993) Environmental responses of plants and ecosystems as predictors of the impact of global change. *Journal of Biosciences*, 18, 4, 515–524.

Chatterjee, N. (1995) Social forestry in environmentally degraded areas of India – case study of the Mayurakshi Basin. *Environmental Conservation*, 22, 1, 20–30.

Cherrett, J.M. (1989a) Key Concepts: The results of a survey of our members' opinions. Chapter 1 in Cherrett, J.M. (ed.) *Ecological Concepts: The contribution of ecology to an understanding of the natural world*. British Ecological Society/Blackwell.

Cherrett, J.M. (ed.) (1989b) *Ecological Concepts: The contribution of ecology to an understanding of the natural world*. British Ecological Society/Blackwell.

Chisholm, M. (1990) The increasing separation of production and consumption. Chapter 6 in Turner, II, B.L., Clark, W.C., Kates, R.W., Richards, J.F., Mathews, J.T. and Meyer, W.B. (1993) *The Earth as Transformed by Human Action: Global and regional changes in the biosphere over the past 300 years*. Cambridge University Press, 87–101.

Christopher, T. and Asher, M. (1996) *The 20-minute gardener: the garden of your dreams without giving up your life, your job, or your sanity*. New York Times, New York.

Clark, W.C. (1989) The human ecology of global change. *International Social Science Journal*, 121 *Reconciling the Sociopshere and the Biosphere*, 315–345.

Clements, F.E. and Shelford, V.E. (1939) *Bio-Ecology*. Wiley, New York.

Cole, E., Erdman, E. and Rothblum, E.D. (1994) Wilderness therapy for women – the power of adventure. *Women and Therapy*, 15, 3–4, 1–8.

Coleman-Cooke, J. (1965) *The Harvest That Kills: An urgent warning about man's use of toxic chemicals on the land*. Odhams.

Collingham, Y.C. and Huntley, B. (2000) Impacts of habitat fragmentation and patch size upon migration rates. *Ecological Applications*, 10, 131–144.

Colten, C.E. (1998) Historical geography and environmental history. *The Geographical Review*, 88, 2, iii–iv.

Cooke, G.W. (1983) *Fertilising for Maximum Yields*. Granada.

Cooper, D.E. and Palmer, J.A. (eds) (1995) *Just Environments: intergenerational, international and interspecies issues*. Routledge.

Corbit, M., Marks, D.L. and Gardescu, S. (1999) Hedgerows as habitat corridors for forest herbs in central New York, USA. *Journal of Ecology*, 87(2), 220–231.

Costanza, R. *et al.* (13 authors) (1997) The value of the world's ecosystem services and natural capital. *Nature*, 387, 15 May.

Courtney, F.M. and Trudgill, S.T. (1984) *The Soil. An introduction to soil study*. Edward Arnold.

Cronon, W. (ed.) (1995a) *Uncommon Ground: Toward reinventing nature*. W. Norton.

Cronon, W. (1995b) The trouble with Wilderness or getting back to the wrong nature. In Cronon, W. (ed.) (1995a) *Uncommon Ground: Toward reinventing nature*. W. Norton.

Cronon, W. (1994) Cutting loose or running aground? *Journal of Historical Geography*, 20, 1, 38–43.

Curl, J.S. (1997) Gardens of allusion. *Interdisciplinary Science Reviews*, 22, 4, 325–342.

Currie, J. (1998) Farmers and their perception of soil conservation methods. In Blume, H.P., Egere, H., Fleischhauer, E., Hebel, A., Reij, C. and Steiner, K.G. (eds) *Towards Sustainable Land Use: Furthering co-operation between people and institutions*. Advances in Geoecology 31, International Society of Soil Science, 1389–1397.

Darier, E. (1999a) *Discourses of the Environment*. Blackwell, 217–140.

Darier, E. (1999b) Foucault against environmental ethics. Chapter 11 in Darier, E. (ed.) *Discourses of the Environment*. Blackwell, 217–240.

Davies, A. (1999) Environmental Education, Ethics and Citizenship; Report of Discussion. *Ethics, Place and Environment*, 2(1), 82–87.

Davies, D. (1988) The evocative symbolism of trees. In Cosgrove, D. and Daniels, S. (eds) *The Iconography of Landscape*, Cambridge University Press, 32–42.

Davis, M.B. (1990) Climatic change and the survival of forest species. Chapter 5 in Woodwell, G.M. *The Earth in Transition: Patterns and processes of biotic impoverishment*. Cambridge University Press, 99–110.

Davis, M. (1998) *Ecology of Fear: Los Angeles and the imagination of disaster*. Metropolitan Books, Henry Holt and Co., New York.

Davison, M. (1998) Experimental design: problems in understanding the dynamical behavior–environment system. *Behavior Analyst*, 21, 2, 219–240.

de Boodt, M. and Gabriels, D. (1980) *Assessment of Erosion*. Wiley.

Defoer, T., Kante, S., and Hilhorst, T. (1998) A participatory action research process to improve soil fertility management. In Blume, H.P., Egere, H., Fleischhauer, E., Hebel, A., Reij, C. and Steiner, K.G. (eds) *Towards Sustainable Land Use: Furthering co-operation between people and institutions*. Advances in Geoecology 31, International Society of Soil Science, 1083–1092.

Dejong, B.H.J., Montoyagomez, G., Nelson, K. and Sotopinto, L. (1995) Community forest management and carbon sequestration – a feasibility study from Chiapas, Mexico. *Interciencia*, 20, 6, 409.

Demeritt, D. (1994) Ecology, objectivity and critique in wrings on nature and human societies. *Journal of Historical Geography*, 20, 1, 22–37.

Demeritt, D. (1996) Social theory and the reconstruction of science and geography *Transactions of the Institute of British Geographers*, 21, 484–503.

Demerrit, D. and Rothman, D. (1999) Figuring the costs of climatic change: an assessment and critique. *Environment and Planning A*, 31, 389–408.

Dewberry, T.C. (1996) *Can we diagnose the health of ecosystems?* Northwest Science and Photography, Florence, Oregon.

Di Chiro, G.D. (1995) Nature as Community: the convergence of environment and social justice. In Cronon, W. (ed.) (1995) *Uncommon Ground: towards reinventing nature*. W. Norton.

Didia, D.O. (1997) Democracy, political instability and tropical deforestation. *Global Environmental Change*, 7, 1, 63–76.

Douthwaite, R. (1998) The best way there: equality and sustainability can't do without each other. *New Internationalist*, 307, 15.

Douglas, I. (1994) Human Settlement. In W.B. Meyer and B.L. Turner II (eds) *Changes in Land Use and Land Use Cover: a global perspective*. Cambridge University Press, pp. 149–169.

Douglas, J S. and de J. Hart, R.A. (1984) *Forest Farming – Towards a solution to problems of world hunger and conservation*. Intermediate Technology Publications, Chapter 5.

Dove, M.R. (1995) The theory of social forestry intervention – the state-of-the-art in Asia. *Agroforestry Systems*, 30, 3, 315–340.

Downing, T.E. (1996) *Climate Change and World Food Security*. NATO.

Draper, M. (1998) Zen and the art of garden province maintenance: the soft intimacy of hard men in the wilderness of KwaZulu-Natal, South Africa, 1952–1997. *Journal of Southern African Studies*, 24, 4, 801–828.

Drury, W.H. (1998) *Chance and Change: Ecology for conservationists*. University of California Press.

Dunbar, R. (1995) *The Trouble with Science*. Faber and Faber.

Eckholm, E. (1979) *The Dispossessed of the Earth: Land reform and sustainable development*. Worldwatch Paper 30, Worldwatch Institute.

Eden, S. (1996) Public participation in environmental policy: considering scientific, counter scientific and non-scientific contributions. *Public Understanding of Science*, 5, 183–204.

Eden, S. (1998) Environmental issues: knowledge, uncertainty and the environment. *Progress in Human Geography*, 22, 3, 425–432.

Eden, S., Tunstall, S.M. and Tapsell, S.M. (1999) Environmental restoration: environmental management or environmental threat? *Area*, 31, 2, 151–159.

Editorial (1996) Honey, it's three AM – What are you doing in the garden? *Forbes*, 106.

Edwards C.A., Lal, R., Madden, P., Miller, R.H. and House, G. (eds) (1990) *Sustainable Agricultural Systems*. Soil and Water Conservation Society.

Elkington, J. and Hailes, J. (1988) *The Green Consumer Guide. From shampoo to champagne – high street shopping for a better environment*. Victor Gollancz.

Ellis, S. and Mellor, A. (1995) *Soils and Environment*. Routledge.

England, K. (1994) From social justice and the city to women friendly cities – feminist theory and politics. *Urban Geography*, 15, 7, 628–643.

Epstein, P.R. (1995) Emerging diseases and ecosystem instability – new threats to public health. *American Journal of Public Health*, 85, 2, 168–172.

Erickson, J. (1990) *Greenhouse Earth: Tomorrow's disaster today*. Tab Books, Blue Ridge Summit, Pennsylvania.

Escobar, A. (1996) Constructing nature: elements for a poststructural political ecology. In Peet, R. and Watts, M. (1996) *Liberation Ecologies: Environment, development, social movements*. Routledge, 46–68.

Evans, D. (1997) *A History of Nature Conservation in Britain*. Routledge.

Evernden, N. (1992) *The Social Creation of Nature*. Johns Hopkins.

Fairhead, J. and Leach, M. (1996) *Misreading the African Landscape: Society and ecology in a forest–savannah mosaic*. Cambridge University Press.

Fankhauser, S. (1995) *Valuing Climatic Change: The economics of the greenhouse*. Earthscan.

Fankhauser, S. and Tol, R.S. (1999) Figuring the cost of climate change: a reply. *Environment and Planning A*, 409–411.

Ffolliott, P.F., Gottfried, G.J. and Rietveld, W.J. (1995) Dryland forestry for sustainable development. *Journal of Arid Environments*, 30, 2, 143–152.

Fiedler, P., White, P.S. and Leidy, R.A. (1997) The paradigm shift in ecology and its implications for conservation. Chapter 6 in Pickett, S.T.A. (ed.) *The Ecological Basis of Conservation: Heterogeneity, ecosystems and biodiversity*. Chapman and Hall.

Fitter, A.H. (ed.) (1985) *Ecological Interactions in Soils*. Blackwell.

Floyd, B. (1970) Agricultural innovation in Jamaica: The Yallahs Valley Land Authority. *Economic Geography*, 46, 1, 63–77.

Forsyth, T. (1996) Science, myth and knowledge: testing Himalayan environmental degradation in Thailand. *Geoforum*, 27, 3, 375–392.

Forsyth, T. (1999) *International Investment and Climate Change*. Royal Institute of International Affairs/Earthscan.

Foster, J. (ed.) (1997) *Valuing the Environment*. Routledge.

Fraser, A.I. (1988) Fuelwood: A review of its global use. *Outlook on Agriculture*, 17, 1, 7–9.

Friday, L. (ed.) (1997) *Wicken Fen: The making of a wetland nature reserve*. Harley Books.

Funtowicz, S.O. and Ravetz, J.R. (1993) Science for the post-normal age. *Futures*, 25, 7, 739–755.

Galbraith, J.K. (1958) *The Affluent Society*. 1962 Pelican edition.

Gallopin, G.C. and Öberg, S. (1992) Quality of life. Chapter 12 in Brennan, M. (compiler). *An Agenda of Science for Environment and Development into the 21st Century*. International Council of Scientific Unions/Cambridge University Press, 227–238.

George, M. (1977) The Norfolk Broads. Transactions of the Norfolk and Norwich Naturalists' Society, 24.

Ghai, D. and Vivian, J.M. (eds) (1992) *Grassroots Environmental Action: People's participation in sustainable development*. Routledge.

Ghazi, P. (1999) Ground Force. *Green Futures*, March/April, 50–51.

Giger, M. (1998) Using incentives and subsidies for sustainable management of agricultural soils – A challenge for projects and policy makers. In Blume, H.P., Egere, H., Fleischhauer, E., Hebel, A., Reij, C. and Steiner, K.G. (eds) *Towards Sustainable Land Use: Furthering co-operation between people and institutions*. Advances in Geoecology 31, International Society of Soil Science, 1405–1411.

Gilbert, O. and Anderson, P. (1998) *Habitat Creation and Repair*. Oxford University Press.

Gilbert, O. (1989) *The Ecology of Urban Habitats*. Chapman and Hall.

Glacken, C.J. (1967) *Traces on the Rhodian Shore: Nature and culture in western thought from ancient times to the end of the eighteenth century*. University of California Press.

Godwin, H. (1975) *A History of the British Flora*. Cambridge University Press.

Goklany, I.M. (1995) Strategies to enhance adaptability – technological change, sustainable growth and free trade. *Climate Change*, 30, 4.

Goldblatt, D. (1996) *Social Theory and the Environment*. Polity Press.

Goldsmith, F.B. and Warren, A. (1993) *Conservation in Perspective*. Wiley.

Golley, F.B. (1993) *A History of the Ecosystem Concept*. Yale University Press.

Goudie, A. (1998) *The Changing Earth: Rates of geomorphological processes*. Blackwell.

Goudie, A. (1999) *The Human Impact on the Natural Environment*. Blackwell.

Goudriaan, J. and Zadoks, J.C. (1995) Global climate-change – modelling the potential responses of agroecosystems with special reference to crop protection. *Environmental Pollution*, 87, 2, 215–224.

Gradwohl, J. and Greenberg, R. (1988) *Saving the Tropical Forests*. Earthscan.

Graf, W.L. (1994) Science, public policy and western American rivers. *Transactions of the Institute of British Geographers*, 17, 5–19.

Greenland, D.J. and Lal, R. (1979) *Soil Conservation and Management in the Humid Tropics*. Wiley.

Greenland D.J. (1981) Soil management and soil degradation. *Journal of Soil Science*, 32, 301–322.

Grey, G. and Deneke, F.J. (1978) *Urban Forestry*. Wiley.

Grimble, R., Chan, M-K., Aglionby, J. and Quan, X. (1995) *Trees and Trade-offs, a stakeholder approach to natural resource management.* Gatekeeper Series, 52, International Institute for Environment and Development.

Grime, J.P. (1979) *Plant Strategies and Vegetation Processes.* Wiley.

Grubb, M., Brack, D. and Vrolijk, C. (1998) *The Kyoto Protocol: A guide and assessment.* Royal Institute of International Affairs/Earthscan.

Hammitt, W.E. and Cole, D.N. (1998) *Wildland Recreation: Ecology and management.* Wiley.

Hamsun, K. (1935) *The Growth of Soil.* 1980 Pan Books edition.

Hannah, L., Lohse, D., Hutchinson, C., Carr, J.L. and Lankerani, A. (1994) A preliminary inventory of human disturbance of world ecosystems. *Ambio*, 23, 4–5, 246–250.

Harding, C. (ed.) (1992) *Wingspan: Inside the Men's Movement.* St Martins Press, New York.

Harrison, C.M. and Burgess, J. (1994) Social constructions of nature: A case study of conflicts over the development of Rainham Marshes. *Transactions of the Institute of British Geographers*, 19, 291–310.

Harrison, P. (1987) *The Greening of Africa: Breaking through in the battle for land and food.* IIED/Earthscan, Paladin Books.

Harvey, D. (1969) *Explanation in Geography.* Edward Arnold.

Harvey, D. (1996) *Justice, Nature and the Geography of Difference.* Blackwell.

Harwood, R.R. (1990) A history of sustainable agriculture. Chapter 1 in Edwards, C.A. *et al.* (eds) (1990) *Sustainable Agricultural Systems.* Soil and Water Conservation Society, 3–19.

Haycock, N., Burt, T.P., Goulding, K.W.T. and Pinay, G. (1997) *Buffer Zones: Their processes and potential in water protection.* Quest International, Harpenden.

Haynes, R.J. (1986) *Mineral Nitrogen in the Plant–Soil System.* Academic Press.

Hayward, T. (1994) *Ecological Thought: An introduction.* Polity Press.

Healey, P. and Shaw, T. (1994) Changing meanings of "environment" in the British Planning system. *Transactions of the Institute of British Geographers*, 19, 425–438.

Henderson, N. (1992) Wilderness and the nature conservation ideal: Britain, Canada and the United States contrasted. *Ambio*, 21, 394–399.

Henderson, C. (1999) Carbon: Soak it up. *Green Futures*, March/April 1999.

Hobley, M. (1996) *Participatory Forestry: The process of change in India and Nepal.* London Overseas Development Institute (Rural Development Forestry Study Guide 3).

Hoffman, A.A. and Parsons, P.A. (1997) *Extreme Environmental Change and Evolution.* Cambridge University Press.

Hoggart, R. (1995) *The Way we Live Now.* Chatto and Windus.

Holland, A., and Rawles, K. (1993) Values in conservation. *ECOS: A Review of Conservation*, 14, 1, 14–19.

Hollick, M. (1993) Self-organizing ecosystems and environmental management. *Environmental Management*, 17, 621–628.

Holling, C.S. (1986) The resilience of terrestrial ecosystems: local surprises and global change. In Clark, W.C. and Munn, R.E. (eds) (1986) *Sustainable Development of the Biosphere.* Cambridge University Press.

Holling, C.S., Peterson, G., Marples, P., Sendzimir, J., Redford, K., Gunderson, L. and Lambert, D. (1996) Self-organization in ecosystems: lumpy geometries, periodicities and morphologies. Chapter 19 in Walker, B. and Steffen, W. *Global Change and Terrestrial Ecosystems.* Cambridge University Press, 346–384.

Holmes, B. (1997) Blazing Chain: if loggers gave up their neat and tidy ways and let rip like a forest fire, nature would reap the benefit. *New Scientist*, 2083, 24 May 1997, 30–33.

Honadle, G. (1993) Institutional constraints on resource use: Lessons from the tropics showing that resource overexploitation is not just an attitude problem and conservation education is not enough. Chapter 3 in Aplet, G.H., Johnson, N., Olson, J. and Sample, V.A. (eds) (1993) *Defining Sustainable Forestry.* The Wilderness Society/Island Press, 90–119.

Hornung M. (1985) Acidification of soils by trees and forests. *Soil Use and Management*, 1, 24–28.

Houghton, J., Meirafilho, L.G., Callender, B.A., Harris, N., Kattenberg, A. and Maskell, K. (eds) (1996a) *Climate Change 1995. The science of climate change.* Cambridge University Press.

Houghton, J. (1997) *Global Warming: The complete briefing* (2nd edn). Cambridge University Press.

Huang, W.H. and Kiang, W.C. (1972) Laboratory dissolution of plagioclase feldspars in water and organic acids at room temperatures. *American Mineralogist*, 57, 1849–1859.

Hugo, V. (1866) *Les Travailleurs de la Mer.* Translation (1911), *Toilers of the Sea.* J.M. Dent.

Huntley, B., Berry, P.M., Cramer, W. and McDonlad, A.P. (1995) Modelling present and potential future ranges of some European higher plants using climate response surfaces. *Journal of Biogeography*, 22, 967–1001.

Huntley, B., Cramer, W., Morgan, A., Prentice, H.C. and Allen, J.R.M. (1997) *Past and Future Rapid Environmental Changes: The spatial and evolutionary responses of terrestrial biota.* NATA ASI Series 1: Global Environmental Change, Vol. 47.

Hurni, H. (1998) A multi-level stakeholder approach to sustainable land management. In Blume, H.P., Egere, H., Fleischhauer, E., Hebel, A., Reij, C. and Steiner, K.G. (eds) *Towards Sustainable Land Use: Furthering co-operation between people and institutions. Advances in Geoecology* 31, International Society of Soil Science, 827–836.

Huston, M.H. (1994) *Biological Diversity: The coexistence of species in changing landscapes.* Cambridge University Press.

Hutchin, P.R., Press, M.C. and Ashenden, T.W. (1996) Methane emission rates from an ombrotrophic mire show marked seasonality which is independent of nitrogen supply and soil temperature. *Atmospheric Environment*, 30, 3011–3015.

Huxley, A. (1932) *Brave New World.* Chatto and Windus. From the Preface, quoted from Granada Publishing edition of 1977 with the 1946 Preface.

Huxley, A. (1937) *Ends and Means: An enquiry into the nature of ideals and into the methods employed for their realization* 1966 Chatto and Windus edition.

Huxley, A. (1962) *Island.* 1976 Granada edition.

Huxley, A. (1977) *The Doors of Perception.* Grafton edition.

Huxley, A. (1997) *The Human Situation.* 1978 Chatto and Windus, 1980 Triad/Granada edition.

Huxley, J.S. (1947) *Conservation of nature in England and Wales.* HMSO.

Ilbery, B., Chiotti, Q. and Rickard, T. (1997) *Agricultural restructuring and Sustainability: A geographical perspective.* CAB International.

Ineson, P., Taylor, K., Harrison, A.F., Poskitt, J., Benham, D.G., Tipping, E. and Woof, C. (1998a) Effects of climatic change on nitrogen dynamics in upland soils. 1. A transplant approach. *Global Change Biology*, 4, 143–152.

Ineson, P., Benham, D.G., Poskitt, J., Harrison, A.F., Taylor, K. and Woods, C. (1998b) Effects of climatic change on nitrogen dynamics in upland soils. 2. A soil warming study. *Global Change Biology*, 4, 153–161.

Innes, J.L. (1991) *Interim Report on Cause-Effect Relationships in Forest Decline*. Geneva. United Nations Economic Commission for Europe.

Irwin, A. (1995) *Citizen Science: A study of people, expertise and sustainable development*. Routledge.

Jeffrey, R. and Sundar, N. (eds) (1998) *A New Moral Economy for India's Forests?* New Delhi: Sage.

Jeffreys, M.V.C. (1962) *Personal Values in the Modern World*. Penguin.

Jepma, C.J. (1995) *Tropical Deforestation: A socio-economic approach*. Earthscan.

Jewitt, S. (1995) Voluntary and "official" forest protection committees in Bihar: solutions to India's deforestation? *Journal of Biogeography*, 22, 1003–1021.

Joad, C.E.M. (1940) *Philosophy for Our Times*. Thomas Nelson.

Johansson, M.B., Berg, B. and Meentemeyer, V. (1995) Litter mass – loss rates in late stages of decomposition in a climatic transect of pine forests – long term decomposition in a Scots Pine forest. *Canadian of Botany*, 73, 1509–1521.

Johns, A.G. (1997) *Timber Production and Biodiversity Conservation in Tropical Rain Forests*. Cambridge University Press.

Johnston, R.J. (1996) *Nature, State and Economy* (especially Chapter 4, Resources, land and environmental use). Wiley.

Jones, M. and Talbot, E. (1995) Coppicing in urban woodlands. *Journal of Practical Ecology and Conservation*, 1, 1, 46–52.

Jordan, C.F. (1998) *Working with Nature: Resource management and sustainability*. Harwood Acdemic Publishers.

Kearns, G. (1998) The virtuous circle of facts and values in the New Western History. *Annals of the Association of American Geographers*, 88, 3, 377–409.

Kekes, J. (1994) Pluralism and the value of life. In Paul, E.F. and Miller, F.F. *Cultural Pluralism and Moral Knowledge*, Cambridge University Press, 44–60.

Kellert, S.R. (1993) The biological basis for human values of nature. In Kellert, S.R. and Wilson, E.O. (1993) *The Bibliophilia Hypothesis*. Island Press, Washington DC.

Kemp, D. (1997) *Global Environmental Issues* (2nd edn). Routledge.

Kendle, T. and Forbes, S. (1997) *Urban Nature Conservation*. Spon.

Keulartz, J. (1998) *The Struggle for Nature: A critique of radical ecology*. Routledge.

Kirilenko, A.P. and Solomon, A.M. (1998) Modelling dynamic vegetation response to rapid climate change using bioclimatic classification. *Climatic Change*, 38, 15–49.

Kirkman, R. (1997) Why ecology cannot be all things to all people. The "adaptive radiation" of scientific concepts. *Environmental Ethics*, 19, 4, 375–390.

Kirschbaum, M.U.F. (ed.) (1995) The temperature-dependence of soil organic-matter decomposition, and the effect of global warming on soil organic-C storage. *Soil Biology and Biochemistry*, 27, 6, 753–760.

Kirschbaum, M.U.F. (ed.) (1996) Ecophysiological, ecological and soil processes in terrestrial ecosystems. In Watson, R.T. *et al.*, *Climate Change 1995: II Impacts, adaptations and mitigation of climate change. Scientific and Technical Analyses*. Cambridge University Press/IPCC, 59–74.

Kononova, M.M. (1966) *Soil Organic Matter*. Pergamon.

Krakauer, J. (1998) *Into the Wild*. Pan Books.

Krott, M. 1996. Self-regulation in forest policy as a challenge to forestry science and practice. Forstwissenschaftliches Centralblatt, 115, 2, 97–107 (English summary from BIDS).

Lackey, R.T. (1998) Seven pillars of ecosystem management. *Landscape and Urban Planning*, 40, 1–3, 21–30.

Lado, C. (1986) Agriculture and environmental knowledge: a case study of peasant farming in Maridi District, Southern Sudan. *Malaysian Journal of Tropical Geography*, 13, 7–36.

Lampkin, N. (1990) *Organic Farming*. Farming Press, Ipswich, UK.

Lantermann, E.D. and Schmitz, B. (1994) Psychic resources and strategies in dealing with global environmental change. Naturwissenschaften, 81, 12, 521–527 (in German, translation of abstract from BIDS).

Latour, B. (1998) To modernise or ecologise? That is the question. Chapter 10 in Braun, B. and Castree N. (eds) *Remaking Reality: Nature at the millennium*. Routledge, 221–242.

Laurie, I.C. (ed.) (1979) *Nature in Cities*. Wiley.

Leakey, R., and Lewin, R. (1995) *The Sixth Extinction: Biodiversity and its survival*. Weidenfield and Nicolson.

Lee, H.S.J. and Jarvis, P.G. (1995) Trees differ from crops and from each other in their responses to increases in CO_2 concentration. *Journal of Biogeography*, 22, 323–330.

Lemons, J. (1996) *Scientific Uncertainty and Environmental Problem Solving*. Blackwell.

Lindenmayer, D.B., Cunningham, R.B. and Pope, M.L. (1999) A large-scale "experiment" to examine the effects of landscape context and habitat fragmentation on mammals. *Biological Conservation*, 88, 3, 387–404.

Lindsay, R. (1995) Galloping Gertie and the precautionary principle; how is environmental impact assessed? Chapter 10 in Wakeford, T. and Walters, M. (1995) *Science for the Earth – can science make the world a better place?* Wiley, 197–236.

Lock, J.M., Friday, L.E. and Bennett, T.J. (1997) The management of the fen. Chapter 11 in Friday, L. (ed.) (1997) *Wicken Fen: The making of a wetland nature reserve*. Harley Books, 213–254.

London, J. (1910) *To Build a Fire*. 1966 Pan Books edition.

Loptson, P. (1995) *Theories of Human Nature*. Broadview Press.

Lovelock, J. (1995) The greening of science. Chapter 3 in Wakeford, T. and Walters, M. *Science for the Earth – can science make the world a better place?* Wiley 39–63.

Luekewille, A. and Wright, R.F. (1997) Experimentally increased soil temperature causes release of nitrogen at a boreal forest catchment in southern Norway. *Global Change Biology*, 3, 13–21.

Luo, Y. and Mooney, H.A. (1999) *Carbon Dioxide and Environmental Stress*. Academic Press.

Luymes, D.T. and Tamminga, K. (1995) Integrating public safety into urban greenways. *Landscape and Urban Planning*, 33, 391–400.

Mabey, R. (1980) *The Common Ground*. Hutchinson.

Mace, G.M., Balmford, A. and Ginsburg, J.R. (1998) *Conservation in a Changing World*. Cambridge University Press.

Mackenzie, A.F.D. (1998) *Land, Ecology and Resistance in Kenya, 1880–1952*. Routledge.

Malitz, J. and Malitz, S. (1998) *Reflecting Nature: Garden designs from wild landscapes*. Bowker Press, New York.

Malone, T.F. (1992) Global change – the human and international dimensions of science – view of the possible. *Interdisciplinary Science Reviews*, 17, 2, 137–141.

Marsh, G.P. (1864) *Man and Nature: or physical geography as modified by human activity*. 1965 edition, Lowenthal, D. (ed.) (1965) Harvard University Press.

Mason, M. (1997) Democratising Nature? The political morality of wilderness preservationists. *Environmental Values*, 6, 281–306.

Massingham, H.J. (1924) *Sanctuaries for Birds and How to Make Them*. G. Bell and Sons.

Mather, A.S. and Needle, C.L. (1999) Development, democracy and forest trends. *Global Environmental Change*, 9, 109–118.

Matthews, E. (1984) Global inventory of pre-agricultural and present biomass. *Progress in Biometeorology*, 3, 237–246.

Matthews, E. (1993) Global vegetation and land use: new high resolution data bases for climate studies. *Journal of Climate & Applied Meteorology*, 22, 474–487.

May, J. (1996) In search of authenticity off and *on* the beaten track. *Environment and Planning, D: Society and Space*, 14, 709–736.

McCann, E., Sullivan, S., Erickson, D. and De Young, R. (1997) Environmental awareness, economic orientation, and farming practices: a comparison of organic and conventional farmers. *Environmental Management*, 21, 5, 747–758.

McCloskey, J.M. and Splading, H. (1989) A reconnaissance-level inventory of the amount of wilderness remaining in the world. *Ambio*, 18, 4, 221–227.

McGuckin, C.P. and Brown, R.D. (1995) A landscape ecological model for wildlife enhancement of stormwater management practices in urban greenways. *Landscape and Urban Planning*, 33, 1–3, 227–246.

McKane, R.B., Rastetter, E.B., Melillo, J. M., Shaver, G.R., Hopkinson, C.S., Fernandes, D.N., Skole, D.L. and Chomentowski, W.H. (1995) Effects of global change on carbon storage in tropical forests of South America. *Global Geochemical Cycles*, 9, 329–350.

Meinig, D.W. (1979) *The Interpretation of Ordinary Landscapes*. Oxford University Press.

Melillo, J.M. *et al.* (5 authors, 29 contributors) (1996) Terrestrial biotic responses to environmental change and feedbacks to climate. Chapter 9 in Houghton, J.F. *et al.* (6 eds). *Climate Change 1995. The science of climate change*. Cambridge University Press, IPCC.

Mendelsohn, R. (1996) An economic-ecological model for ecosystem management. In Adamowicz, W.L., Boxall, P.C., Luckert, M.K., Philips, W.E. and White, W.A. (eds) (1996) *Forestry, Economics and Environment*. CAB, 213–221.

Merchant, C. (1980) *The Death of Nature: Women, ecology and the scientific revolution*. Harper, San Francisco.

Meyer, W.B. and Turner II, B. L. (eds) (1994) *Changes in Land Use and Land Use Cover: A global perspective*. Cambridge University Press.

Milchunas, D.G. and Lauenroth, W.K. (1995) Inertia in plant community structure – state changes after cessation of nutrient enrichment stress. *Ecological Applications*, 5, 2, 452–458.

Miller, G. T. (1988) *Living in the Environment*. Wadsworth, Belmont, California.

Minteer, B.A. and Manning, R.E. (2000) Convergence in environmental values: an empirical and conceptual defence. *Ethics, Place and Environment*, 3, 47–60.

Mitchell, T.D. and Hulme, M. (1999) Predicting regional climate change: living with uncertainty. *Progress in Physical Geography*, 23, 1, 57–78.

Mollison, B. C. and Holmgren, D. (1978) *Permaculture: A perennial agricultural system for human settlements*. Corgi Books.

Mooney, H.A., Cushman, J.H., Medina, E., Sala, O.E. and Schulze, E.F. (1996) *Functional Roles of Biodiversity. A global perspective*. SCOPE 55, Wiley.

Morgan R.P.C. (1980) Soil erosion and its control in Britain. *Progress in Physical Geography*, 4, 24–47.

Morgan R.P.C. (1986) *Soil Erosion and Conservation*. Longman.

Muhs, D.R. (1984) Intrinsic thresholds in soil systems. *Physical Geography*, 5, 2, 99–110.

Munda, G. (1997) Environmental economics, ecological economics and the concept of sustainable development. *Environmental Values*, 6, 213–233.

Napier, T.L. (1990) The evolution of US soil conservation policy: from voluntary action to coercion. Chapter 43 in Boardman, J., Foster, I.D.L. and Deering, J.A., *Soil Erosion on Agricultural Land*. Wiley, 627–644.

Nasar, J.L., Fisher, B. and Gronnis, M. (1993) Proximate cues to fear of crime. *Landscape and Urban Planning*, 26, 1–4, 161–178.

Nash, R. (1973) *Wilderness and the American Mind*. Yale University Press.

Newton, R.G. (1997) *The Truth of Science*. Harvard University Press.

Nicholson, E.M. (1929) *How Birds Live*. Williams and Norgate.

Nicholson, M. (1957) *Britain's Nature Reserves*. Country Life Limited.

Nicholson, M. (1970) *The Environmental Revolution: A guide for the new masters of the world*. Pelican.

Nicholson-Lord, D. (1987) *The Greening of the Cities*. Routledge.

Norton, B. (1955) Applied philosophy v. practical philosophy: towards an environmental philosophy integrated according to scale. In Marietta, D. and Imbree, L. (eds) *Environmental Philosophy and Environmental Activism*. Rowman and Littlefield, Baltimore, 125–148.

Oakley, P. and Marsden, D. (1984) *Approaches to participation in rural development*. International Labour Organisation (International Labour Office, CH-1211, Geneva 22, Switzerland).

Odum, E.P. (1978) *The value of wetlands: a hierarchical approach*. American Water Resources Association: Proceedings of the National Symposium on Wetlands.

Odum, E.P. (1989) *Fundamentals of Ecology*. Sanders.

OECD (1995) *Sustainable Agriculture*. Organization for Economic Co-operation and Development.

Oeschlager, M. (1991) *The Idea of Wilderness from Prehistory to the Age of Ecology*. Yale University Press.

Olwig, K. (1996) Nature – mapping the ghostly traces of a concept. Chapter 3 In Earle, C., Mathewson, K. and Kenzer, M.S. *Concepts in Human geography*, Rowan and Littlefield, London, 63–96.

O'Neill, O. (1997) Environmental values, anthropocentrism and speciesism. *Environmental Values*, 6, 127–142.

Opie, J. (1998) Moral geography in high plains history. *The Geographical Review*, 88, 2, 241–258.

Orians, G.H. (1975) Diversity, stability and maturity in ecosystems. In van Dobben, W.H. and Lowe-MConnell, R.H. (eds) *Unifying Concepts in Ecology*. D.W. Junk, pp. 139–149.

O'Riordan, T. (1976) *Environmentalism*. Pion Books.

O'Riordan, T. (1992) Introduction: Responses and Strategies in Brennan, M. (compiler). *An Agenda of Science for Environment and Development into the 21st Century*. International Council of Scientific Unions/Cambridge University Press, 223–225.

O'Riordan, T. (ed.) (1995) *Environmental Science for Environmental Management*. Longman.

O'Riordan, T. (1997) Valuation as revelation and reconciliation. *Environmental Values*, 6, 169–183.

O'Riordan, T. and Jordan, A. (1999) Institutions, climate change and cultural theory: towards a common analytical framework. *Global Environmental Change*, 9, 81–93.

Osler, M. (1990) A Word about Boxes. *Hortus*, No. 16.

Osterkamp, W.R. and Hupp, C.R. (1996) The evolution of geomorphology, ecology and other composite sciences. In Rhoads, B.L. and Thorn, C.E. (eds) *The Scientific Nature of Geomorphology*. Wiley.

Owens, S. (1994) Land, limits and sustainability: a conceptual framework and some dilemmas for the planning system. *Transactions of the Institute of British Geographers*, 19, 439–456.

Pacione, M. (1999) Applied geography: in pursuit of useful knowledge. *Applied Geography*, Vol. 19, 1–12.

Pahl-Wostl, C. (1995) *The Dynamic Nature of Ecosystems: Chaos and order entwined*. Wiley.

Palmer, J. (1998a) *Environmental Education for the 21st Century*. Routledge.

Palmer, J. (1988b) Spiritual ideas, environmental concerns and educational practice. In D.E. Cooper and J.A. Palmer, *Spirit of the Environment: Religion, value and environmental concern*. Routledge, pp. 146–167.

Park, C.C. (1992) *The Tropical Rain Forest*. Routledge.

Pattanayak, S. and Mercer, E. (1998) Valuing soil conservation benefits of agroforestry: contour hedgerows in the Eastern Visayas, Philippines. *Agricultural Economics*, 18, 31–46.

Patten, B. (1971) *The Irrelevant Song*. George Allen and Unwin.

Pearce, D., Barbier, E. and Markandya, A. (1990) *Sustainable Development: Economics and environment in the third world*. Earthscan.

Pearce, D., Markandya, A and Barbier, E. (1989) *Blueprint for a Green Economy*. Earthscan.

Pepper, D. (1993) *Eco-socialism: from deep ecology to social justice*. Routledge.

Perlman, D.L. and Adelson, G. (1997) *Biodiversity: Exploring values and priorities in conservation*. Blackwell.

Peterken, G.F. (1996) *Natural Woodland: Ecology and conservation in northern temperate regions*. Cambridge University Press.

Phifer, P. (1999) Review of Grumbine, R.E., *Ghost Bears: Exploring the Biodiversity Crisis, Ethics, Place and Environment*, 2, 1, 121–122.

Pickett, S.T.A. and White, P.S. (1985) *The Ecology of natural disturbance and patch dynamics*. Academic Press.

Pimm, S.L. (1991) *The Balance of Nature? Ecological issues in the conservation of species and communities*. Chicago University Press.

Plamondon, W.M. (1996) Energy and Leadership, Chapter 28 in Hesselbein, F., Goldsmith, M. and Beckhard, R. (1996). *The Leader of the Future: New visions, strategies and practices for the next era*. Jossey-Bass Publishers, San Francisco.

Pomeroy, L.R. and Alberts, J.L. (eds) (1986) *Concepts of Ecosystem Ecology: A comparative view*. New York.

Poore, D. (1989) *No Timber Without Trees: Sustainability in the tropical forest*. Earthscan.

Porritt, J. (1984) *Seeing Green: The politics of ecology explained*. Blackwell.

Porritt, J. and Winner, D. (1988) *The Coming of the Greens*. Fontana.

Porteous, J.D. (1996) *Environmental Aesthetics: Ideas, politics and planning*. Routledge.

Powch, I.G. (1994) Women and Therapy – what makes it empowering for women. *Wilderness Therapy*, 15, 3–4, 11–27.

Pratt, B. and Boyden, J. (1985) *A Field Director's Handbook: An Oxfam manual for development workers*. Oxfam.

Pretty, J. (1998) Furthering cooperation between people and institutions. In Blume, H.P., Egere, H., Fleischhauer, E., Hebel, A., Reij, C. and Steiner, K.G. (eds) *Towards Sustainable Land Use: Furthering co-operation between people and institutions*. Advances in Geoecology 31, International Society of Soil Science, 837–850.

Proctor, J.D. (1995) Whose Nature? The contested moral terrain of ancient forests. In Cronon, W. (ed.) (1995a) *Uncommon Ground: Toward reinventing nature*. W. Norton.

Proctor, J.D. (1998a) The social construction of nature: relativist accusations, pragmatist and critical realist responses. *Annals of the Association of American Geographers*, 88, 3, 352–376.

Proctor, J.D. (1998b) Geography, paradox and environmental ethics. *Progress in Human Geography*, 22, 2, 234–255.

Proctor, J.D. and Pincetl, S. (1996) Nature and the reproduction of endangered space: the spotted owl in the Pacific Northwest and southern California. *Environment and Planning D: Society and Space*, 14, 683–708.

Quigley, P. (1999) Nature as dangerous space. Chapter 9 in Darier, E. (ed.) (1999a) *Discourses of the Environment*. Blackwell, 182–202.

Rackham, O. (1986) *A History of the British Countryside*. Dent.

Rawles, K. (1998) Philosophy and the environmental movement. Chapter 10 in Cooper, D.E. and Palmer, J. *Spirit of the Environment: Religion, value and environmental concern*. Routledge, 131–145.

Rebele, F. (1994) Urban ecology and special features of urban ecosystems. *Global Ecology and Biogeography*, Letters, 4, 6, 173–187.

Rees, W. and Wackernagel, M. (1994) Ecological footprints and appropriated carrying capacity: measuring the natural capital requirements of the human economy. In Jansson, A-M. *et al.* (eds) *Investing in Natural Capital*. Island Press, Washington.

Reganold, J.P. (1988) Comparison of soil properties as influenced by organic and conventional farming systems. *American Journal of Alternative Agriculture*, 3, 144–155.

Reganold, J.P. (1989) Farming's organic future. *New Scientist*, 10 June, 49–52.

Reganold, J.P., Elliot, L.F. and Unger, Y.L. (1987) Long-term effects of organic and conventional farming on soil erosion. *Nature*, 330, 370–372.

Reij, C., Scoones, I. and Toulmin, C. (1996) *Sustaining the Soil: Indigenous soil and water conservation in Africa*. Earthscan.

Reij, C. (1998) How to increase the adoption of improved land management practices by farmers? In Blume, H.P., Egere, H., Fleischhauer, E., Hebel, A., Reij, C. and Steiner, K.G. (eds) *Towards Sustainable Land Use: Furthering co-operation between people and institutions*. Advances in Geoecology 31, International Society of Soil Science, 1413–1420.

Rejmanek, M. (1999) Holocene invasions: finally the resolutions ecologists were waiting for! *Trends in Evolutionary Ecology*, 14, 1, 8–10.

Retallack, G.J. (1990) *Soils of the Past: An introduction to paleopedology*. Unwin Hyman.

Retallack, G.J. (1992) Paleozoic paleosols. In Martini, I.P. and Chesworth, W. *Weathering, Soil, and Paleosols*. Developments in Earth Surface Processes, 2. Elsevier, 543–564.

Retallack, G.J. (1996) Palaeosols: Record and engine of past global change. *Geotimes*, June 1996, 25–28.

Rhodes, F.H.T. (1962) *The Evolution of Life*. Pelican.

Roberts, J.M. (1996) *The Penguin History of Europe*. Penguin.

Robinson, A.B., Baliunas, S.L., Soon, W. and Robinson, Z.W. (1998) Environmental effects of increased carbon dioxide. *Medical Sentinel*, 3(5), 171–178.

Rocheleau, D.E. (1995) Environment, development, crisis and crusade – Ukambane, Kenya. *World Development*, 23, 6, 1037–1051.

Rockwell, T.K., Johnson, D.L., Keller, E.A. and Dembroff, G.R. (1985) A late Pleistocene–Holocene soil chronosequence in the Ventura basin, southern California, USA. Chapter 16 in Richards, K.S., Arnett, R.R. and Ellis, S. (eds) *Geomorphology and Soils*. George Allen and Unwin, 308–327.

Röling, N.G. and Wagemakers, M.A.E. (1998) *Facilitating Sustainable Agriculture: Participatory learning and adaptive management in times of environmental uncertainty.* Cambridge University Press.

Rolston, II, H. (1997) Nature, the genesis of value and human understanding. *Environmental Values*, 6, 361–364, and see Discussion 355–360.

Rothschild, M. and Marren, P. (1997) *Rothschild's Reserves: Time and fragile nature.* Balaban Harley.

Rowell, A. (1996) *Green Backlash: Global subversion of the environmental movement.* Routledge.

Royal Commission on Environmental Pollution, Report 19 (1996) The *Sustainable Use of Soil*. Cm 3165, HMSO; see also under its Chairman: Houghton (1966).

Russell, B. (1940) *An Inquiry into Meaning and Truth.* 1962 Pelican edition.

Russell, E.J. (1943) Agriculture in Europe after the War: What can we do? *Journal of the Royal Agricultural Society of England*, 104, 151–163.

Russell, E.J. (1961) *Soil Conditions and Plant Growth.* Wiley.

Sack, R.D. (1990). The realm of meaning: the inadequacy of human–nature theory and the view of mass consumption. Chapter 40 in Turner, II, B.L., Clark, W.C., Kates, R.W., Richards, J.F., Mathews, J.T. and Meyer, W.B. (1990) *The Earth as Transformed by Human Action: Global and regional changes in the biosphere over the past 300 years.* Cambridge University Press, 659–671.

Saint-Exupery, A. de (1939.) *Wind, Sand and Stars.* 1966 Penguin edition.

Samson, P.R. and Pitt, D. (eds) (1999) *The Biosphere and Noosphere Reader: Global environment, society and change.* Routledge.

Sanders, D., Theerawong, S. and Sombatpanit, S. (1998) Soil conservation extension: from concepts to adoption. In Blume, H.P., Egere, H., Fleischhauer, E., Hebel, A., Reij, C. and Steiner, K.G. (eds) *Towards Sustainable Land Use: Furthering co-operation between people and institutions.* Advances in Geoecology 31, International Society of Soil Science, 1421–1428.

Schama, S. (1996) *Landscape and Memory.* Fontana.

Schlesinger, M.E. (ed.) (1991) *Greenhouse Gas Induced Climatic Change: A critical appraisal of simulation and observations.* Elsevier.

Schwarz, W. and Schwarz, D. (1987) *Breaking Through: the theory of holistic thinking.* Green Books.

Scoones, I. (1997) The dynamics of soil fertility change: historical perspectives on environmental transformation from Zimbabwe. *The Geographical Journal*, 163, 2, 161–169.

Seidman, S. (1998) *Contested Knowledge: Social theory in the post-modern era.* Blackwell.

Shafer, C.L. (1990) *Nature Reserves: Island theory and conservation practice.* Smithsonian Institution.

Shafer, C.S. and Hammitt, W.E. (1995) Congruency among experience dimensions, condition indicators, and coping behaviours in wilderness. *Leisure Sciences*, 17, 4, 263–279.

Shapin, S. (1998) Placing the view from nowhere: historical and sociological problems in the location of science. *Transactions of the Institute of British Geographers*, 23, 5–12.

Sharma, R.A. (1993) The socioeconomic evaluation of social forestry policy in India. *Ambio*, 22, 4, 219–224.

Sheail, J. (1976) *Nature in Trust*. Blackie.

Sheail, J. (1995) The ecologist and environmental history – a British perspective. *Journal of Biogeography*, 22, 953–966.

Sheail, J. (1987) *Seventy Five Years in Ecology: The British Ecological Society*. Blackwell.

Shrader-Frechette, K. and McCoy, E. (1993) *Method in Ecology: Strategies for conservation*. Cambridge University Press.

Shugart, H.H. (1998) *Terrestrial Ecosystems in Changing Environments*. Cambridge University Press.

Simmons, I.G. (1982) *Biogeographical Processes*. George Allen and Unwin.

Simmons, I.G. (1993a) *Environmental History: a concise introduction*. Blackwell.

Simmons, I.G. (1993b) *Interpreting Nature: Cultural constructions of the environment*. Routledge.

Singh, B., El Maayar, M., Andre, P., Bryant, C. R. and Thouez, J-P. (1998) Impacts of GHG-induced climate change on crop yields: effects of acceleration in maturation, moisture stress and optimal temperature. *Climate Change*, 38, 51–86.

Sjöberg, L. (1989) Global change and human action: psychological perspectives. *International Social Science Journal: Reconciling the Sociosphere and the Biosphere*, 121, 413–432.

Sloan, L.C. and Rea, D.K. (1996) Atmospheric carbon dioxide and early Eocene climate: A general circulation modelling sensitivity study. *Palaeogeography, Palaeoclimatology, Palaeoecology*, 119, 3–4, 275–292.

Smales, J. (1999) Pleasure gardens and productivity. *Green Futures*, March/April, 37.

Smith, D. (1997) Back to the good life: towards an enlarged concept of social justice. *Environment and Planning D: Society and Space*, 15, 19–35.

Smith, D.I. (1998) *Water in Australia: Resources and Management*. Oxford University Press.

Smith, M., Whitelegg, J. and Williams, N. (1998) *Greening the Built Environment*. Earthscan.

Smith, T.B., Bruford, M.W. and Wayne, R.K. (1993) The preservation of process: the missing element of conservation programs. *Biodiversity Letters*, 1, 164–167.

Smith, V.K. (1995) Does education induce people to improve the environment? *Journal of Policy Analysis and Management*, 14, 4, 599–604.

Solecki, W.D. and Welch, J.M. (1995) Urban parks – green spaces or green walls. *Landscape and Urban Planning*, 32, 2, 93 106.

Soper, M.H.R. and Carter, E.S. (1985) *Modern Farming and the Countryside: The issues in perspective*. Association of Agriculture.

Steinbeck, J. (1939) *The Grapes of Wrath*. 1995 Mandarin Paperbacks edition.

Steinbeck, J. (1951) *The Log from the Sea of Cortez*. 1960 Pan Books edition.

Stevenson, R.L. (1879). *Travels with a Donkey in the Cevennes*. Kegan Paul.

Stocking, M. (1998) Making development sustainable. In Blume, H.P., Egere, H., Fleischhauer, F., Hebel, A., Reij, C. and Steiner, K.G. (eds) *Towards Sustainable Land Use: Furthering co-operation between people and institutions*. Advances in Geoecology 31, International Society of Soil Science, 851–856.

Storr, A. (1989) *Solitude: A return to the self*. Fontana.

Strutt, N. (1970) *Modern Farming and the Soil*. HMSO.

Sullivan, S., McCann, E., DeYoung, R. and Eickson, D. (1996) Farmers attitudes about farming and the environment – a survey of conventional and organic farmers. *Journal of Agricultural and Environmental Ethics*, 9, 2, 123–143.

Sulzman, E.W., Poiani, K.A. and Kittel, T.G.F. (1995) Modelling Human Induced Climatic Change – A Summary for Environmental Managers. *Environmental Management*, 19, 2, 197–224.

Sutherland, W.J. (ed.) (1998) *Conservation Science and Action*. Blackwell.

Sutherland, W.J. and Hill, D.A. (1995) *Managing Habitats for Conservation*. Cambridge University Press.

Syers, J.K. and Rimmer, D.L. (eds) (1994) *Soil Science and Sustainable Land Management in the Tropics*. CAB International/British Society of Soil Science.

Tansley, A.G. (1935) The use and abuse of vegetational concepts and terms. *Ecology*, 16, 284–307.

Teughels, I., Van Hecke, P. and Impens, I. (1995) Competetion in a global change environment: the importance of different plant traits for competetive success. *Journal of Biogeography*, 22, 297–305.

Teysott, G. (ed.) (1999) *The American Lawn*. Princeton Architectural Press.

Thayer, R.L. (1976) Visual ecology: revitalizing the aesthetics of landscape architecture. *Landscape*, 20, 37–43.

Thoreau, H.D. (1854) *Walden* and (1906) *Journal* and other quotes from *In Wilderness is the Preservation of the World*. 1967 Sierra Club/Ballantine Books edition.

Thornley, J.H.M. and Cannell, M.G.R. (1997) Temperate grassland responses to climate change: an analysis using the Hurley Pasture Model. *Annals of Botany*, 80, 205–221.

Tiffen, M., Mortimore, M. and Gichuki, F. (1994) *More People, Less Erosion? Environmental recovery in Kenya*. Wiley.

Timberlake, L. *et al.* (1993) Forest principles. In *Agenda 21. Facing the Future*. Earthscan.

Tinker, P.B. (1988) The efficiency of the agriculture industry in relation to the environment. Chapter 1 in Park, J.R. *Environmental Management in Agriculture*. Belhaven Press.

Tinker, I. (1994) Women and community forestry in Nepal – expectations and realities. *Society and Natural Resources*, 7, 4, 367–381.

Trudgill S.T. (1978) "The earth will endure and blossom forth in spring." *MIMS Magazine*, 3, 15 September, 249–258.

Trudgill, S.T. (1988) *Soil and Vegetation Systems*. Oxford University Press.

Trudgill, S.T. (1989) The Government and the environment: An interview with Nicholas Ridley. *Geography Review*, 3(1), 15–18.

Trudgill, S.T. (1990) *Barriers to a better environment: what stops us solving environmental problems?* Belhaven/Wiley.

Trudgill, S.T. (1999) Environmental Education, Ethics and Citizen Conference held at the Royal Geographical Society (with the Institute of British Geographers), 20 May 1998. Introduction, and Postscript. *Ethics, Place and Environment*, 2, 1, 81–82, 87–89.

Trudgill, S.T., Chell, K. and Riley, C. (1996) Education and conservation issues in the Slapton Ley NNR. *Field Studies*, 8, 7272–746.

Trudgill, S.T. and Coles, N. (1985) The movement of nitrate fertiliser from the soil surface to drainage waters by preferential flow in weakly structured soils, Slapton, S. Devon. *Agriculture, Ecosystems and Environment*, 13, 241–259.

Trudgill, S.T. and Richards, K.S. (1997) Environmental science and policy: generalizations and context sensitivity. *Transactions of the Institute of British Geographers*, 22, 5–12.

Trudgill, S.T., Walling, D. and Webb, B. (1999) *Water Quality: Processes and policy.* Wiley.

Turner, B.L. (1994) Local faces, global flows – the role of land use and land cover in global environmental change. *Land Degradation and Rehabilitation*, 5, 2, 71–78.

Turner, T. (1995) Greenways, blueways, skyways and other ways to a better London. *Landscape and Urban Planning*, 33, 1–3, 269–282.

Tyler, G.A., Smith, K.W. and Burges, D.J. (1998) Reedbed management and breeding bitterns *Botarus stellaris* in the UK. *Biological Conservation*, 86, 2, 257–266.

UK Round Table on Sustainable Development (1998) *Aspects of Sustainable Agriculture and Rural Policy.* UK Round Table on Sustainable Development.

Ulrich, R.S. (1984) View through a window influences recovery from surgery. *Science*, 224, 420–421.

Unwin, R.J. (1998) Farmer perception of soil protection issues in England and Wales. In Blume, H.P., Egere, H., Fleischhauer, E., Hebel, A., Reij, C. and Steiner, K.G. (eds) (1998). *Towards Sustainable Land Use: Furthering co-operation between people and institutions.* Advances in Geoecology 31, International Society of Soil Science, 1399–1404.

USDA (1964) *Land of Washington.* US Department of Agriculture Soil Conservation Service, Washington.

Usher, M.B. (1973) *Biological Management and Conservation.* Chapman and Hall.

Usher, M.B. and Williamson, M.H. (1974) *Ecological Stability.* Chapman and Hall.

Uzzell, D.L., Rutland, A. and Whinstance, D. (1995) Questioning values in environmental education. Chapter 12 in Guerrier, Y. *et al.* (eds) *Values and the Environment: A social science perspective.* John Wiley.

van der Post, L. (1974) *A Far-Off Place.* Penguin.

van der Post, L. (1978) *Jung and the Story of Our Time.* Penguin.

van der Post, L. (1986) *A Walk with a White Bushman.* Penguin.

van Dobben, W.H. and Lowe-McConnell, R.H., (eds) (1975) *Unifying Concepts in Ecology.* Dr W. Junk B.V. Publishers.

van Dyne, G.M. (1969) *The Ecosystem Concept in Natural Resource Management.* New York.

Van Matre, S. (1983) *The Earth Speaks.* Acclimatization Experiences Institute, PO Box 288 Warrenville, Illinois, 60555.

VEMAP Members (1995) Vegetation/ecosystem modelling and analysis project: Comparing biogeography and biogeochemistry models in a continental-scale study of terrestrial ecosystem responses to climate change and CO_2 doubling. *Global Biogeochemistry Cycles*, 9, 4, 407–437.

Vinal, William ("Cap'n Bill") Gould (1940) *Nature Recreation: Group guidance for the out-of-doors.* 1963 Dover Publications edition.

Vira, B., Dubois, O., Daniels, S.E. and Walker, G.B. (1998) Institutional pluralism in forestry: considerations of analytical and operational tools. *Unasylva*, 49, 35–42.

Vitousek, P.M. (1994) Beyond global warming – ecology and global change. *Ecology*, 75, 7, 1861–1876.

Vohra, B.B. (1985) *The Greening of India.* Indian National Trust for Art and Cultural Heritage, New Delhi, INTACH Environmental Series, 1.

Von Weizsacker, E., Lovins, A.B. and Lovins, L.H. (1997) *Factor Four: Doubling wealth, halving resource use.* The new report to the Club of Rome. Earthscan.

Wakeford, T. and Walters, M. (1995) *Science for the Earth – Can science make the world a better place?* Wiley.

Walker, B. and Steffen, W. (1996) *Global Change and Terrestrial Ecosystems*. Cambridge University Press.

Walker, B.H. (1994) Landscape to regional-scale responses of terrestrial ecosystem to global change. *Ambio*, 23, 1, 67–73.

Walker, D. (1989) Diversity and Stability. Chapter 5 in Cherrett, J.M. (ed.) *Ecological Concepts: The contribution of ecology to an understanding of the natural world*. British Ecological Society/Blackwell, 115–145.

Walmsley, A. (1995) Greenways and the making of urban form. *Landscape and Urban Planning*, 33, 1–3, 81–127.

Walton, D.W. and Bridgewater, P. (1996) Of gardens and gardeners. *Nature and Resources*, 32, 3, 15–19.

Wang, Y.P. and Polglase, P.J. (1995) Carbon balance in the Tundra, Boreal Forest and Humid Tropical Forest during climatic change – scaling up from leaf physiology and soil carbon dynamics. *Plant Cell and Environment*, 18(10), 1226–1244.

Warren, A. (1995) Changing understanding of African pastoralism and the nature of environmental paradigms. *Transactions of the Institute of British Geographers*, 20, 193–203.

Warrick, R.A., Gifford, R.M. and Parry, M.L. (1986) CO_2, climatic change and agriculture. Chapter 9 in Bolin, B., Doos, B.R., Jager, J. and Warrick, R.A. (1986) *The Greenhouse Effect, Climatic Change and Ecosystems*. Wiley.

Watt, K.E.F. (1973) Principles of Environmental Science. McGraw-Hill.

Watts, W.A. (1988) Europe. In Huntley, B. and Webb. III, T. (eds) *Vegetation History*, Kluwer Academic Publishers, Dordrecht, 155–192.

Webb, III, T. (1986) Past changes in vegetation and climate: lessons for the future. In Peters, R.L. and Lovejoy, T.E. (eds), *Global Warming and Biological Diversity*. Yale University Press, 59–75.

Weischet, W. and Caviedes, C.N. (1993) *The Persisting Ecological Constraints of Tropical Agriculture*. Longman.

Westhoff, V. (1983) Man's attitude towards vegetation. In Holzner, W., Wenger, M.J.A. and Ikusima, I. (eds). *Man's Impact on Vegetation*. D.W. Junk, The Hague.

Wheeler, D. (ed.) (1998) *The Penguin Book of Garden Writing*. Penguin.

Whitmore, T. (1998) *An Introduction to Tropical Rain Forests*. Cambridge University Press.

Whitney, G.G. (1994) *From Coastal Wilderness to Fruited Plain: A History of environmental change in temperate North America from 1500 to the present*. Cambridge University Press.

Whittaker, J.B. and Tribe, N.P. (1998) Predicting numbers of an insect (*Neophilaenus lineatus*: Homptera) in a changing climate. *Journal of Animal Ecology*, 67, 987–991.

Whittaker, R.H. (1975) *Communities and Ecosystems*. Macmillan.

Wiens, J.R. (1984) On understanding a non-equilibrium world: myth and reality in community patterns and processes. In Strong, D.R. *et al.* (eds) *Ecological Communities: Conceptual issues and the evidence*. Princeton.

Wiersum, K.F. (1995) 200 years of sustainability in forestry – lessons from history. *Environmental Management*, 19, 3, 321–329.

Wilby, R.L. (1995) Greenhouse hydrology. *Progress in Physical Geography*, 19, 3, 351–369.

Williams, M. (1994) The relations of environmental history and historical geography. *Journal of Historical Geography*, 20, 1, 3–21.

Williams, D.R. and Patterson, M.E. (1996) Environmental meaning and ecosystem management: perspectives from environmental psychology and human geography. *Society and Natural Resources*, 9, 5, 507–521.

Williamson, H. (1922) *The Lone Swallows and other essays of boyhood and youth.* 1945 Putnam edition.

Wilson, A. (1992) *The Culture of Nature: North American landscapes from Disney to the Exxon Valdez.* Blackwell.

Wilson, G.A. and Bryant, R.L. (1997) *Environmental Management: New directions for the twenty-first century.* UCL Press.

Wiman, L.B. (1991) Implications of environmental complexity for science and policy. *Global Environmental Change*, 1, 235–247.

Winteringham, F.P.W. (1985) *Environment and Chemicals in Agriculture.* Elsevier.

Woodward, F.I., Smith, T.M. and Emanuel, W.R. (1995) A global land primary productivity and phytogeography model. *Global Biogeochemical Cycles*, 9, 4, 471–490.

Woodwell, G.M. (ed.) (1990) *The Earth in Transition. Patterns and processes of biotic impoverishment.* Cambridge University Press.

Worster, D. (1994) *Nature's Economy: A history of ecological ideas.* Cambridge University Press.

World Resources Institute (1988) *World Resources 1988–89.* Basic Books, New York.

Wyatt, J. (1973) *The Shining Levels: The story of a man who went back to Nature.* Geoffrey Bles.

Yaalon, D. and Berkowicz, S. (eds) (1997) *History of Soil Science – International perspectives.* Advances in Geoecology, 29. Reiskirchen: Catena Verlag.

Young, A. (1989) *Agroforestry for Soil Conservation.* CAB International.

Young, A. (1998) *Land Resources: Now and for the future.* Cambridge University Press.

Zelinsky, W. (1975) The Demigod's dilemma. *Annals of the Association of American Geographers*, 65, 2, 123–143.

Zimmerer, K.S. (1996a) Ecology as cornerstone and chimera in human geography. Chapter 6 in Earle, C., Mathewson, K. and Kenzer, M.S. *Concepts in Human Geography*, Rowan and Littlefield.

Zimmerer, K.S. (1996b) Discourses on Soil Loss in Bolivia: Sustainability and the search for the socioenvironmental "middle ground". Chapter 5 in Peet, R. and Watts, M. (1996) *Liberation Ecologies: Environment, development, social movements.* Routledge, 110–124.

Zimmerman, M.E. (1994) *Contesting Earth's Future: Radical ecology and postmodernity.* University of California Press.

Zola, E. (1880) *La Terre.* Translation (1980) *The Earth.* Penguin.

Subject index

Figures: ***bold italics***; Tables and insets: **bold**

Author index